THE

POETICAL GEOGRAPHY,

WITH THE

RULES OF ARITHMETIC IN VERSE.

[View of Edfau.]

The towns and mountains which beneath me stood,
And rivers rolling to the dark-blue flood,
And isles and lakes as they were spread to me,
I 'll sing and bind upon thy memory:
Harsh sounds in smooth unbroken lines shall glide
As free and easy as the sparkling tide.

PUBLISHED AT CINCINNATI.

PRICE TWENTY-FIVE CENTS.

Walter. M. Collins Bok

THE

POETICAL GEOGRAPHY,

DESIGNED TO ACCOMPANY

OUTLINE MAPS OR SCHOOL ATLASES.

TO WHICH ARE ADDED THE RULES OF ARITHMETIC IN RHYME.

BY GEORGE VAN WATERS.

[Black and White Swan.]

The towns and mountains which beneath me stood,
And rivers rolling to the dark-blue flood,
And isles and lakes as they were spread to me,
I 'll sing and bind upon thy memory:
Harsh sounds in smooth unbroken lines shall glide
As free and easy as the sparkling tide.

PUBLISHED AT CINCINNATI.

AND SOLD BY AGENTS ONLY.

1852.

To those who have not the time to turn over a large volume, who are not prejudiced against improvements in science and literature, who would learn the leading features of a very difficult branch and keep it in the memory, who have a relish for the novel, and a curiosity to know some of the most important and striking characteristics of nature and art; in a word, all unprejudiced and generous minds, to such, this work is respectfully dedicated and inscribed by their

Humble Servant,

THE AUTHOR,

who has spared neither time nor labor to make it useful and interesting.

PREFACE.

One of the extravagances of authors, is, in flattering themselves that their own productions are superior to those of their rivals, and many, to make it appear more obvious than it may be, essay to turn the public favor from their competitor's merits, by defects real or pretended, which they are careful to exhibit; presenting their own at the same time, in their most brilliant colors.

But, avoiding this extreme, the Author of this work would ask leave only to show the need and worth of a rhyming system, in securing a knowledge of the branch here treated of, and leave others to decide how far he has succeeded in the formation of such.

That proper names are more difficult to retain in the memory, than common, is almost superfluous to mention. In acquiring general terms, or in the study of language, the continual repetition of the same words, and our familiarity with the subjects to which they are applied, renders it less difficult to the memory, than in learning proper names that are fixed to denote one thing only, and never occur unless the objects or things for which they stand are particularized.

Geography is a branch that is studied by nearly all—but how few among the vast number, who spend years in acquiring a knowledge of it, ever retain or remember it.

But the defect is not to be attributed to the works studied, but to the poverty of memory; the retentive powers of the mind are not endowed with energies competent to the task of sustaining so cumbrous a load; some mechanical aid is requisite, and hence the utility of a work of the present kind.

The author has endeavored to circumscribe in as small a space as possible, the matter here presented, and by so doing, has sacrificed ornament to brevity, which is the leading characteristic of the work.

From the different pronunciations that names admit of, and the unsettled difference among the learned and unlearned, as to their correct orthoepy, the manner of pronunciation here, may by many, be deemed imperfect, and by a different pronunciation, render many of the lines prosaic and disproportionate in measure, which will doubtless be an objection offered to the reception of the work; but such an objection would be unjust and unwarrantable; it would be utterly impossible to establish a system of orthoepy, which would be sanctioned by all. That of the present work is founded principally on the authority of Baldwin, Worcester, and Morse—and if theirs be adhered to, no irregularities of sound or quantity will be discernible.

The following, from Joseph E. Worcester, will serve to show the impossibility of establishing a uniform system of pronunciation of foreign names, and also, the high claims of Common Custom (and it might be said with propriety, "Common Sense"), in settling this matter.

"There can be no doubt but that geographical names, which assume such different forms in different languages, should be pronounced differently by the inhabitants of different countries, and in accordance with the analogies of their respective languages. All the common geographical names, such as are familiar to all intelligent persons, have become more or less Anglicized, and their pronunciation is more or less conformed to the English analogy. Many of these words may be considered as perfectly Anglicized, and are pronounced as common English words; but there are many that are only partially Anglicized, and with regard to such, it is often difficult to determine how far, in pronouncing them, the English analogy should be allowed to prevail.

"With respect to the class of words which are partially Anglicized, there is a great diversity in the manner of pronouncing them. Some respectable speakers incline to pronounce them, for the most part, according to the English analogy, while others aspire to pronounce them as they are pronounced in the several languages to which they appertain; and there are many cases in which it is difficult to determine which is most to be approved, the English or foreign method; but a medium between the two extremes may be regarded generally as a judicious course."

San Francisco in 1848.

San Francisco in 1849.

INTRODUCTION.

AWAY into a grove young Alva strayed,
His task to learn beneath the cooling shade;
Before him lay an Atlas open wide,
Where towns and mountains stood on every side;
Long on its page his studious mind was placed,
But dark Forgetfulness each name defaced;
At length discouraged, sorrow o'er him press'd,
And a deep sigh came from his laboring breast,
When lo! a seraph stood before his face,
And beamed with radiance of celestial grace;
In his right hand a golden lyre he held,
And 'mid ambrosial clouds poised o'er the field;
The astonished boy could scarce his presence brook,
While the fair Spirit thus his errand spoke:

"I am a traveler, on my ærial way,
Across the gulf of vast immensity
I speed my course, and in a moment pass,
From star to star—from world to universe.
Creation's furthest skirts I have beheld,
And marshalled o'er her wide unbounded field;
And when I winged the vast profound of space,
This world remote reared up her clayey face;
With rapid flight, upon extended oars
I came and circled round her terrene shores—
All I beheld—but ere I passed away
To other worlds, I cast mine eyes on thee.
I saw the tear roll from thy sparkling eye,
And why it rolled, I need not ask thee why;

I 've come, my boy, to wipe the falling tears,
And give an opiate for thy grief and fears:
The towns and mountains which beneath me stood,
And rivers rolling to the dark-blue flood,
And isles and lakes as they were spread to me,
I 'll sing and bind upon thy memory:
Harsh sounds in smooth unbroken lines shall glide
As free and easy as the sparkling tide.
When first I launched me down the ethereal sky,
Columbia's shores were spread before mine eye
In dusky features, for the orb of day
Blazed on the antipodes, the other way,
And the pale moon, fair empress of the night,
Sat veiled in glory, on her chariot bright.
And now of that, which lay beneath my wing,
Harp, raise thine airs, and aid me as I sing."
Thus having said, he struck his heavenly lyre,
And sang in accents of celestial fire.

THE

POETICAL GEOGRAPHY.

I.—GEOGRAPHICAL DEFINITIONS.

Geography.

The surface of the Earth, with all its tribes,
Of sea and land, Geography describes.

The Earth.

This Earth is but a mighty ball profound,
Just five and twenty thousand miles around:
One fourth the surface of this globe is land;
Three fourths are water, as you understand.

Prose Definitions.

GEOGRAPHY—Geography is a description of the earth's surface.

The earth is a large ball, the diameter of which is eight thousand miles, and the circumference, or distance around it, twenty-five thousand.

One fourth of the surface of the earth is land, and the other three fourths water.

The earth is one of the planets that revolve around the sun; which circuit it performs once in a year. It turns round upon its axis once in twenty four hours. Its distance from the sun is ninety five millions of miles.

II.—DIVISIONS OF LAND.

Divisions.

Of land, and its divisions, read the stories;
Peninsulas, *Continents*, *Islands*, *Promontories*,
And *Isthmuses* and *Capes*, and *Mountains* high,
Volcanoes, *Shores*, and *Deserts*, wet and dry.

The Land is divided into Continents, Islands, Promontories, Isthmuses, Capes, Mountains, Volcanoes, Shores, Deserts, &c

The surface of land, to the surface of water, is 1 to 4; but the cubic proportions are unknown.

Continents.

A *Continent* is a vast extent of land,
Where rivers run and boundless plains expand;
Where mountains rise—where towns and cities grow,
And nations live, and all their care bestow.
Two continents only, on this globe are seen—
Eastern and *Western*, are their names (I ween);
The *Eastern Continent*, we see, divide
In *Europe*, *Africa*, and *Asia* wide.
The *Western Continent* we next behold,
Where *North* and *South America* unfold.

A Continent is a vast extent of land, not divided by water.

There are two continents—the Eastern and Western. The Eastern Continent comprises Europe, Asia, and Africa. The Western Continent comprises North and South America.

Islands.

Islands, upon all sides, the waves surround;
In rivers, lakes, and seas, and oceans found.

An Island is a portion of land, surrounded by water; as, Long Island, Isle of France, Isle of Man, Iceland, Ireland, &c.

Peninsulas.

A Peninsula, the dark sea wave entwines,
Save by some neck that to the main land joins.

A narrow portion of land, extending into the sea, is called a Peninsula; as, Malacca, California, &c.

Mountains.

Mountains are high and elevated land,
That rises o'er the province, dark and grand.

A Mountain is a high elevation of land, that rises above the surrounding country; as, Mount Sinai, Mount Holyoke, the White Mountains, &c. The top of a mountain is called the summit; the bottom is the foot, or base. When the land rises to a small hight, it is called a *hill*. The space between two hills or mountains is called a valley.

When the land is flat and level, it is called a plain. Extensive plains are called, in the United States, prairies; as, Rock Prairie, in Rock county, Wisconsin. In South America, they are called pampas or llanas. In Asia, steppes; as, the Steppes of Issim, in the southwestern part of Siberia.

Valleys.

Valleys are spaces 'tween the mountains spread,
Safe from the storm that scathes the mountain's head.

Valleys are spaces between mountains, or hills. They are sometimes called vales.

Volcanoes.

Volcanoes, from their craters, vomit fire,
And smoke and lava, in a stream, most dire.

Volcanoes are mountains that send forth fire and smoke from their tops, and sometimes melted stones. The opening in the top, is called the *crater*. The discharge of melted matter, is called an *eruption*. The matter thrown out, is called *lava*.

Capes.

A point of land extending in the sea,
Is called a Cape; as Cape Romania.

Promontories.

When high above the waves, or dark seas hoary,
The proud Cape hangs, 't is called a *Promontory*.

A Cape is a point of land extending into the sea; as Cape Horn, Cape Ann, &c.

A high Cape is a Promontory.

Deserts.

A Desert is a vast and sandy plain,
Where sweeps the simoom and the hurricane,
Where vegetation neither grows nor thrives,
Where nothing finds repose, and no one lives.

A Desert is a vast and sandy plain, destitute of vegetation; as Sahara in Africa, Attacama in South America.

A fertile spot in a Desert is called an Oasis; as the Oasis of Fezzan in Sahara. These Oases in the Deserts serve as resting places for caravans that cross them.

III.—DIVISIONS OF WATER.

The Water next, of this great globe we mention,
Of *Seas* and *Oceans* first, of vast extension,
Then *Archipelagoes*, and *Gulfs*, and *Bays*,
And *Lakes* and *Channels*, next the eye surveys,
And *Sounds*, and *Friths*, and *Roads*, and *Harbors* too,
With *Rivers* rolling to the dark seas blue.

The Water is divided into Oceans, Seas, Gulfs, Bays, Archipelagoes, Lakes, Channels, Straits, Harbors, Roads, Havens, &c.

Oceans.

An *Ocean* is a vast extent of brine,
Or salt sea water, boundless and sublime.

An Ocean is a vast extent of salt water not separated by land; as the Atlantic and Pacific Ocean.

The Ocean goes by different names, as the Main, Sea, Deep, Brine, &c.

Seas.

Seas are large bodies of the briny tide,
By land encircled round on every side.

A Sea is a collection of salt water surrounded by land; as the Caspian Sea.

Archipelagoes.

A Sea filled full of Islands, well you know,
Is always called an *Archipelago*.

A Sea filled full of Islands is called an Archipelago; as the Grecian Archipelago.

Gulfs, or Bays.

A Gulf or Bay, is when the waves expand
To wide extent, encroaching on the land.

When the sea, or water extends up into the land, it is called a Gulf or Bay; as the Bay of Fundy, Gulf of Bothnia, &c.

Lakes.

Lakes are fresh water Seas, and always found,
By land compassed upon all sides around.

A Lake is a body of water surrounded by land, the same as a sea, only that the water is fresh instead of salt; as Lake Erie.

Straits.

A narrow passage, like a door or gate,
That leads into some sea, is called a *Strait.*

A passage of water that leads between two seas, or bodies of water, is called a Strait; as the Straits of Magellan, between South America and the Island of Terra del Fuego.

Channels.

A Channel is a strait that opens wide;
As the *English Channel,* where proud navies ride.

A Channel is a wide strait; as the English Channel.

Sounds.

A Strait so shallow that its depth is found,
By lead or anchor, oft is called a sound.

When a strait is so shallow that its depth can be measured by a lead and line, it is called a *Sound.*

Rivers.

Rivers are streams, by numerous branches formed,
That from the highlands to the seas are turned.

A River is a large stream of water, formed by numerous branches, that empties into some sea, gulf, lake or bay. The place where a river rises, is called its source; the place where it empties is called its mouth. The small streams that empty into it are called its branches.

Firths.

A River wid'ning 'tween its banks of earth,
Towards its mouth, is called a *Frith or Firth.*

The widening of a river toward its mouth, is called a Frith or Firth; as Solway Frith in Scotland; the Firth of the River Forth.

Harbors or Havens.

A Harbor or a Haven, is a port,
Where ships in safety, from the storm resort.

A Harbor or Haven is a port where ships may run in and find shelter from the storm.

EXPLANATIONS NECESSARY TO THE USE OF MAPS.

Hemispheres.

The world's a Globe, the world we live on here;
One half a globe is called a *Hemisphere.*

Eastern and *Western Hemispheres* are found
Upon the Map that shows, the world is round.
Northern and *Southern Hemispheres* beside,
One North, one South the Equator is espied.

The word *hemisphere* is formed from *hemi*, that signfies half, and *sphere*, globe or ball; so, half the earth is called a *hemisphere.*

The Western Hemisphere includes North and South America.

The Eastern Hemisphere includes Europe, Asia and Africa.

The Northern Hemisphere includes all that part of the earth North of the Equator.

The Southern Hemisphere includes all South of the Equator.

The Equator.

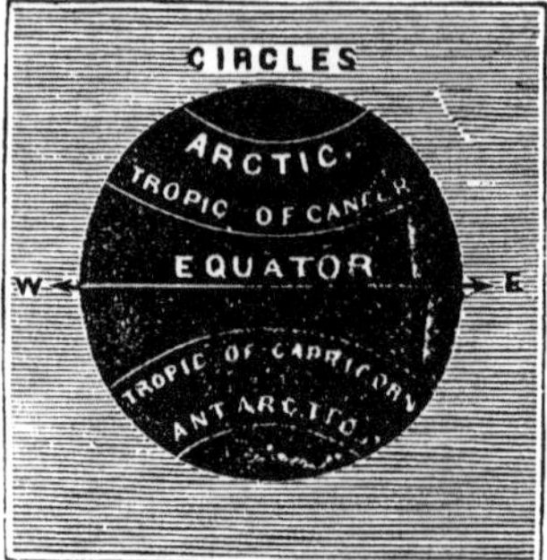

A circle drawn around the earth, and greater
Than any parallel, is called the *Equator.*
The *Equator* is a fancied line, that folds
Around the earth, half way between the poles.
This circle 's called the *Equinoctial Line,*
For when the Solar orb doth o'er it shine,
The days and nights are equal, as the clocks
And watches all proclaim the *Equinox.*

That part of the earth which is just half way between the North and South poles, or equally distant from the poles, is called the Equator. It is the warmest part of the Earth, as the sun's rays are more direct on this portion than any other. It is called by mariners simply, the *Line.*

Tropics.

Tropics are circles that restrict the sun,
Which with the equator parallel doth run,
Just *twenty-three and a half* degrees they shine
Both North and South the Equatorial line.

The North is *Cancer* called, South *Capricorn,*
For here the sun doth in his pathway turn,
And backward trace his steps; these circles show
The limits of the solar orb below.

Tropics are circles that run parallel with the Equator, at the distance of twenty three and a half degrees North and South of it. The circle North of the Equator, is called the *Tropic of Cancer.* The one South of the Equator, is called the *Tropic of Capricorn.*

Tropic signifies return; for when the sun arrives as far from the Equator as either of these lines, it appears to stop and retrace its steps.

The sun crosses the Equator twice a year; on the 21st of March and the 21st of September.

It is over the Tropic of Cancer the 21st of June; which is called the Summer Solstice. This is the longest day in the year, to all North of the Equator, and the shortest to all South of it.

The sun is over the Tropic of Capricorn the 21st of December; this is called the Winter Solstice. It is the shortest day in the year, to all North of the Equator, and the longest to all South of it.

Polar Circles.

And of the Polar Circles now I'll tell:
They with the *Tropics* are found *parallel;*
Just *twenty three, one half,* and nothing less ($23\frac{1}{2}$),
Aloof the Poles;—these, in degrees, I guess.

The Polar Circles are parallel with the Tropics, and $23\frac{1}{2}$ degrees from the Poles. This, in geographic miles, would be 1410 miles, the distance from the Pole to the Circle; twice this distance, or 2820 miles, is the diameter of the Arctic or Antarctic Circle, or the Frigid Zones.

When the Sun is over the tropic of Cancer, all that part within the Arctic Circle has constant day; and all that part in the tropic of Capricorn, constant night. The reverse takes place when the Sun is over the tropic of Capricorn, on the 21st of December.

At the Poles it is day six months of the year, without intermission; for this length of time the sun is visible above the horizon. The other six months of the year, it is one dark, dreary night.

Meridians.

Meridians run from *Pole* to *Pole* ('tis true),
Cutting the Equator, at right angles, through;
They're used to reckon distance, east and west,
And of all other ways have proved the best.

Parallels of Latitude.

Now Parallels of Latitude we'll view:
They are lines that pass around the globe (not through),
As parallel they with the Equator run,
Eastward and westward is the course they turn.

Meridians are, also, imaginary lines, drawn on the Map, to reckon distance, east or west, from any one of them. They run from the North to the South Pole.

All places through which the same meridian passes have noon, or midnight, at the same time.

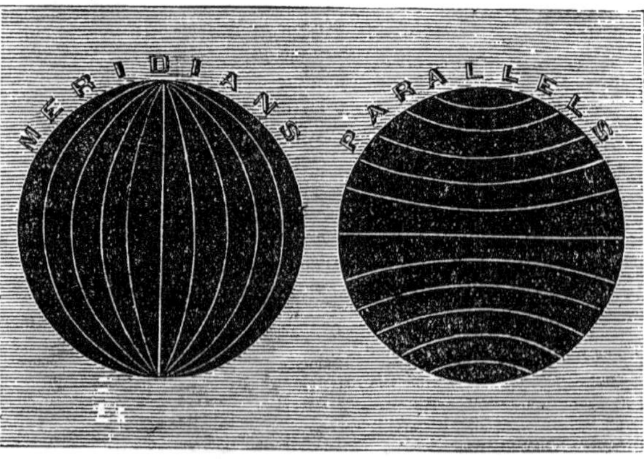

Parallels of Latitude are lines on the Map, used to reckon distances, north or south, of the Equator.

Latitude and Longitude.

Latitude is distance from the Equator,
Either north or south (let it be less or greater);
The distance, east or west, is *Longitude*
From any one meridian, understood.
Both *latitude* and *longitude* are reckoned
In *minutes* and *degrees*, as well as *seconds;*
Just sixty geographic miles make a degree;
In English, sixty nine and just a half you see.

Places that are under the Equator have no Latitude. All places north of the Equator are in *North Latitude;* all places south of the Equator are in *South Latitude.* The greatest latitude a place can have is 90 degrees. The North and South Poles of the earth are the only places that have 90 degrees of latitude.

Longitude is reckoned east and west.

Most nations reckon their longitude from the metropolis of their country; as, the French, from Paris; the English, from Greenwich; the Americans, from Washington. Though the Americans reckon mostly from Greenwich, the same as the English.

Latitude and longitude are reckoned in degrees, minutes, and seconds. Sixty geographic miles (or sixty nine and a half English miles), make a degree; sixty seconds make a minute; sixty minutes one degree.

Every circle is supposed to be divided into 360 degrees, whether it be larger or smaller.

The distance round the Earth being 360 degrees, one half of that distance must be 180 degrees; one quarter, 90 degrees.

The greatest distance that any two objects on the surface of the earth can be apart, is 180 degrees. To be this distance, they must be on opposite sides of the earth; consequently, no place can have over 180 degrees of longitude.

Zones.

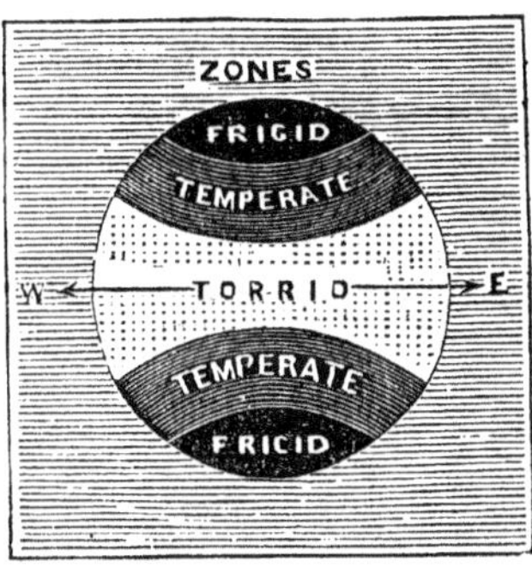

Zones.

Zones are divisions of earth's surface; made
By *tropics* and the *polar circle's* aid.
There are *five zones:* two *temperate* and one *torrid,*
Two *frigid zones,* where winter's cold is horrid.

Torrid Zone.

The Torrid Zone is 'tween the tropics spread,
Where, twice a year, the sun is overhead.

Temperate Zones.

The Temperate Zones are on those parts our ball,
Which 'tween the polar curves and tropics fall.

Frigid Zones.

The Frigid Zones include both land sky,
Of parts which in the polar circles lie.

The Torrid, or Burning Zone, is included within the Tropics.

The Temperate Zones are between the Tropics and the Polar Circles.

The Frigid Zones include those parts of the earth between the Polar Circles and the Poles.

The Temperate Zones enjoy a mild, temperate climate, between the extremes of heat and cold.

The heat in the Torrid, or Burning Zone, is excessive at all seasons of the year.

In the Frigid, or Frozen Zone, the cold is intense. Winter holds an uninterrupted reign the year round.

Maps.

A Map's a picture, of the whole or part,
Of the earth's surface, to be learned by heart.
The top is North, while South points to your breast;
The right hand's East, the left hand's always West.
More Maps than one, bound up for school or college,
Is called an *Atlas,* and contains much knowledge.

How Latitude and Longitude are expressed on Maps.

Both Latitude and Longitude, we see,
Upon the Map, in figures, 1—2—3;
Upon the side the Latitude is told,
While Longitude we at the top behold.

Maps are pictures of the whole, or of parts, of the earth's surface. The top of the map represents the northern part of a country; the bottom, the southern; the right hand, the eastern; the left hand, the western. A collection of maps is called an *Atlas.*

Latitude and Longitude are represented on Maps by figures. Latitude is usually written on the sides of the map, while Longitude is generally at the top or the bottom.

QUESTIONS are not inserted in this work, from the fact that it was deemed superfluous. All the Teacher has to do, to form a question, is to read over any sentence and prefix the interrogatives, 'What is —?' or, 'Where are —?' &c., and it becomes a question.

He turns to page 7, for instance, and glances his eye upon the word, Geography, in full-faced letters, (over the poetry that defines Geography,) and asks the question—'What is Geography?' Then, 'What is the Earth,' &c.; and, to answer the question, the Pupil repeats the poetry, and in his own language gives the sum and substance of the prose.

The Teacher, in all instances, should explain the licensed poetic phrases to juvenile classes.

NORTH AMERICA.

North America is noted for the largest lakes of fresh water in the world, and as being the home of the oppressed of all nations.

View of Niagara Falls from the Ferry.

Capes.

Cape Farewell, south of Greenland, first relate,
While *Wel'-sing-ham* is west of Davis Strait,
Cape Lewis stands southeast of Es'-qui-maux,
And North of Bell'-isle Strait, as seamen know
North of the Gulf, and South of Lab'-ra-dor,
Cape Whittle hears Law-ren'-e-an surges roar.
Cape Sable west, by Nova Scotia's formed,
Where Fundy's matchless *tide* is backward turned.
Then comes *Cape Ann, Cape Cod* and *Mal-a-bar',*
Of Massachusetts all, as you're aware.
Rhode Island holds *Point Ju'-dith, Point Mon-tauk',*
Long Island claims, and Jersey *Sandy Hook,*
Cape Hen'-lopen of Delaware—*Cape May,*
Of Jersey's shore, by Delaware's dark Bay.
Just at the entrance of the Ches-a-peake',
Cape Charles and *Henry* both their sentries keep.
Cape Hatteras, then *Cape Lookout* and *Cape Fear,*
Of North Carolina, in the list appear.
Of Florida, *Can-ave-ral,* well you know,
With *Florida, Sable* and *Ro'-ma-no,*
And one *St. Blas,* near Ap-pa-lach'-ee's flow,
Ro-xo' beside Tam-pi'-co next appears; (Ro-ho.)
Ca-touche' of Yu-ca-tan' the sailor nears. (Ca-toosh.)
East of Honduras *Gra'-cias* mantles low,
As west of Cuba stands *An-to'-ni-o.*
As the Western Coast of Mexico we keep,
First *Co-ri-ents'* springs upward from the deep,
St. Lucas next, and *St. La-za'-ro's* seen,
Mor-ro Her-mo'-so then, and *Point Mon-drains'.*
Men-do'-ci-no o'er *forty* latitude, (40)
While *Oxford Cape,* for *forty three* is good. (43)
In *sixty-five,* and east of *Beh'-ring's* pass, (65)
Cape Prince of Wales, heaves up his icy mass;
Just North of this, *Cape Lisbon* you behold,
Then *Icy Cape* and *Bar'-rows Point* unfold;
And *Bee'-cher, De-mar-ca-tion, Bath'-urst,* all
Where the frozen ocean scours the arctic wall.

PRONUNCIATION.

Esquimeaux,	*Es-ke-mo.*	Mondraines,	*Mon-drene.*
Tampico,	*Tam-pe'-co.*	Henlopen,	*Hen'-lo-pen.*
Catouche,	*Ca-toosh.*	Roxo,	*Ro-ho'.*

Rivers.

Cold Ar-a-bas'-ca Lake, the *Elk* divides,
And the *Peace River,* to *Slave River* guides.
Slave River, to Slave Lake her tribute pays,
And to a Northern Strait, *McKenzie* strays.
The *Seal* in Hudson's ample Bay is rolled,
With *Churchill, Nelson* and the *Severn* cold.
And *Albany* and *Moose,* in James' Bay,
With *East Maine River* all their waters lay.
Red River ends in *Winnipeg* her story,
Where *Sas-ka-shaw'-an* mantles all her glory.

RIVERS THAT CONNECT THE LAKES, &c.

St. Mary's from Superior, Huron takes,
In Lake St. Clair, *St. Clair* from Huron breaks.
From Lake St. Clair, *Detroit* to Erie pours,
From whence *Niagara* to Ontario roars:
From here *St. Lawrence* to the gulf drives in,
With *U-ta-was,* her first and largest stream.

RIVERS ON THE EASTERN COAST.

St. John's from Maine, through Brunswick makes her way,
And with *St. Croix,* rolls into Fundy's Bay. (St. Croy)
From Maine, *Penobscot,* and the *Ken-ne-beck,*
With *An-dros-cog-gin* and the *Sa-co* break.
The *Mer-ri-mack,* from Hampshire takes its coil
Through Massachusetts' northern, eastern soil.
Connecticut, in the Eastern States is found,
With *Hou-sa-ton-ic* wakes Long Island Sound.
Mohawk to Hudson, *Hudson* to the sea,
From New York State, join in the Jubilee.
From Jersey's shore, the *Delaware* divides
The Delaware and Pennsylvanian meads.
By Maryland and Pennsylvania formed,
In *Ches-a-peake,* the *Sus-que-han-nah's* turned.
Here *Po-to-mac* drives onward to the brine,
'Tween Maryland and Virginia the line.
Here *Rhap-pa-han-nock, York* and *James* are thrown
From fair *Virginia,* their summer home.
In Al-be-marle, *Cho-wan,* and *Roanoke,*
Virginia's shores with Carolina yoke.

And *Pam-li-co* and *Neuse* in Pamlico,
O'er North Carolina, murmur in their flow,
From whence *Cape Fear* to Ocean mutters low.
Little and *Great Pe-dee'*, here take their source,
And with *San-tee'*, through South Carolina course.
Edis-to here, with *Cam-ba-hee* entwine,
With dark *Sa-van-nah* on the Georgian line.
O-gee-chee, *Al-ta-ma-ha*, *Satilla*, all
From Georgia drive, and into ocean fall;
From whence *St. Mary's* waves to Ocean stray,
With *Nas-sau* and *St. John's* in Florida.

RIVERS OF THE GULF OF MEXICO.

From Georgia, *Flint* and *Chat-a-hoo'-chee* lower,—
The *Chatahoochee* bounds her Western shore.
Then to the Gulf, o'er Florida they stray,
Through *Ap-pa-lach-i-co'-la's* watery way.
Mo-bile from *Alabama* comes, whose bed
The *Alabama* and *Tom-big'-bee* tread.
And *Pas-ca-gou-la* and the limpid *Pearl*,
From Mississippi State, their waters furl,
And *Mississippi* here unloads her stores,
And the broad Gulf her boiling surge devours.
'Tween Louisiana and the Texan green,
Through *Lake Sa-bine'*, here rolls the dark *Sa-bine'*,
From Texas, *Ne'-ches*, *Trin-i-dad*, and *Brazos*,
With *Col-o-ra'-do* whose loud roar would *craze us*.
Here *Gau-da-loupe'* and *Nue-ces* ceaseless flow,
With *Rio Grande*, northeast of Mexico.

MISSISSIPPI.

The *Mississippi*, from I-tas'-ca Lake
In I'-o-wa,* bids the broad Gulf awake.
Wisconsin for the Eastern Coast survey,
Then Illinois, Kentucky, Tennessee.
Then Mississippi's soil is next beheld,
With Louisiana's most southeastern field
With I'-o-wa; Missouri's on the west.
Where, with Arkansas, Louisiana's pressed.

EASTERN BRANCHES.

To this proud tide, from broad Wisconsin, flock
The *Chip-pe-wa*, *Wisconsin*, and the *Rock*.
Through Illinois, *Rock river* rolls her tide,
Where *Illinois* and fair *Kas-kas'-kia* glide.
Ohio, here, from Pennsylvania comes;
South of Ohio state her billow foams.
Then Indiana state, and Illinois, (*illinoy*)
Beside its pathway all their arts employ.
The same dark breaker sweeps Virginian earth,
And bounds Kentucky state upon the north.
In Western Tennessee, the *O-bi-on* keeps,
And *Hatch-ie* on to Mississippi sweeps;
In Mississippi state, behold *Ya-zoo'*,
In zigzag path, with *Big Black river*, too

WESTERN BRANCHES.

From I'-o-wa, the *Willows*, and the *Pine*,
Crow Wing, and *Swan*, and *Elk*, and *Sack*, combine;
Then, casting up their bubbles by the billion,
Crow river comes, *St. Peters*, and *Vermillion*.
White Water, *Root*, and *Upper Iowa*,
With *Turkey river*, sing their roundelay.

* Sometimes, though erroneously, pronounced I-o'-wa.

Red Cedar then, with *Iowa* made fast;
Skunk river next with dark *Des Moines* the last.

MISSOURI.

Far from Missouri Territory driven,
Where the Rocky Mountains rear their heads to heaven,
Leaving fair Iowa upon the east,
Missouri rolls to Mississippi's breast

And *Mar-a-mec* springs from the Iron Mountain,
And runs northeast, to *Mississippi's* fountain.
Arkansas state lays claim to young *St. Francis*,
Where, from the Rocky Mountains, foams *Ar-kan-sas*.*
The branches of the last are, *White* and *Jean*;
Both in Arkansas, on the map are seen.
And, to *Arkansas*, with the tide *Neo-sho*,
From Indian fields, *Canadian* murmurs low.
O'er Texas, north, southwest the Arkansan banner,
Red river, comes; then pours through Louisiana.

BRANCHES OF THE OHIO.

'Tween Illinois and Indiana, tread
The *Wa-bash* billows, to *Ohio's* bed.
White river, with her *East* and *Western* tides,
From Indiana, to the *Wabash* glides.
Ohio state hears, with *Mi-a-mi's* roar
Scioto, *Hocking* and *Musk-ing-um*—four,
Then *Beaver river*, born in Pennsylvania,
Last northern branch, save one, called *Al-le-gha-ny*.
Ohio drinks *Mo-non-ga-he-la* in,
That sweeps Virginia and the land of Penn
Little Kan-ha-way, then, is on the route,
With *Great Kanhaway* and the *Guy-an-dotte*.
Big Sandy, on Virginia's western border;
And *Licking river*, of Kentuckian order.
Then comes *Kentucky river*, *Salt*, and *Green*—
Upon the last, the Mammoth Cave is seen.
Where Tennessee, and state Kentucky, slumbers,
The *Cum-ber-land* warbles her watery numbers.
In the same states, and Alabama, too,
The *Tennessee* unfolds unto the view.

BRANCHES OF THE MISSOURI.

Among Missouri's branches, on the north,
Are *Thompson's*, *Williams'*, *Porcupine*, *White Earth*,
From Iowa, the *Shepherd* rolls her tides,
With *Fish*, and *James*, and *Sioux*, and *river Floyds*,
With *Nod-a-way*, from state Missouri, run
The *Platte*, and *Grande*, with *river Char-a-ton*.
The *Jefferson* a northern course is thrown,
To join the *Madison* and *Yellowstone*.
The *Yellowstone* collects, in her own sea,
Clark Fork, *Big Horn*, and *Tongue*—of branches three.
Little Missouri next, then *Cannon Ball*,
Chay-enne, and *Platte*, all from Missouri fall.
La-platte is on her territorial bound,
North of the Indian claims and hunting ground,
With branches four—*Big Horn*, *Loup Fork*, and *Black*,
With one *Pa-dou-ca*, on the Indian tract.
And eastward, from the Indian Province, wide
The *river Kanzas* rolls her three-forked tide.

* Sometimes pronounced Ar-kan-saw'.

The northern fork, *Republican*, behold;
Then *Sol-y-man's*, and *Smoky Hill*, unfold.
La-mine and fair *O-sage*, pour forth their waters,
With *Gasconudé*, Missouri's warbling daughters.

RIVERS OF MICHIGAN.

From Michigan, the *Raisin* runs to Erie,
And *Huron*, there her sparkling waters carry,
In *St. Clair River*, ***Gratiot*** ends her lay,
And ***Sag-i-naw*** is lost in her own bay;
With branches ***Cass***, and ***Flint***, and ***Shi-a-was-see***,
Which roar, at last with ***Huron's*** billows, massy.
And ***Tit-ti-ba-was-see***, with her ***Chippewa***,
That drinks the ***Pine***, is lost in *Saginaw*.
Au-sa-ble eastward runs, with *Thunder Bay*,
Where *Huron's* billows greet them on the way.
Che-boy-gan, northward, leaves the noisy clan,
While *Betsey*, westward, seeks lake Michigan;
There, *Min-is-tee*, and *Not-i-pe-ka-go*, run,
With *river White*, and flaming *Mus-ke-gon*.
Grand river, then, and the dark *Kal-ma-zoo*,
With one *St. Josephs*, break their pathway through.

RIVERS OF NORTHERN NEW YORK.

The *Sa-ra-nac* is lost in lake Champlain,
East of the state, where Plattsburgh holds her reign;
St. Reg-is, *Racket*, and the *river Grass*,
With *Os-we-gatch-ie*, to *St. Lawrence* pass.
Black river, then, *Oswego*, *Genessee*,
Ontario drowns in her loud minstrelsy.

RIVERS OF OREGON AND CALIFORNIA.

In Oregon, *Columbia* makes her pillow,
And westward, to Pacific, drives her billow.
From California, *Sacramento*'s roll'd;
Southward her course, through regions rich with gold.
And *Colorado* joins with *Gila river*;
In *California Gulf*, their waves roll ever.

PRONUNCIATION.

Chippewa, -	*Chippewau'.*	Multnomah,	*Mult-no'-ma.*
Mobile, - -	*Mobeel'.*	Sioux, - -	*Soo.*
Sabine, - -	*Sabeen'*	Chayenne, -	*Sha-en'.*
Itasca, - -	*I-tas'-ca.*	Laplatte, -	*La-plate'.*
Hatchie, -	*Hach'-ee.*	Lamine, -	*La-meen'.*
Des Moines,	*De Moin.*	Saline, - -	*Sa-leen.*
St. Croix, -	*St. Croy.*	Hiaqui, - -	*He-a-kee'.*
Gila, - -	*Ge-la*, or *Hee-la.*	Sascashawan,	*Sas-ca-shaw'-an.*

COUNTRIES OF NORTH AMERICA.

THE RUSSIAN POSSESSIONS

Are noted for their furs, and as a cold, dreary climate, inhabited only by savages and hunters. The population is about 50,000.

The coast, in some parts, rises into snow capped summits.

GREENLAND.

The coldest, most dreary, and desolate country in the world. Inhabited by a filthy, degraded race of Indians, called Esquimaux.

Captain Ross, on the northern shores, found a race of ignorant natives, who imagined his ships to be huge birds. On the cliffs, he found red snow.

Esquimeaux spearing a Walrus.

The settlements are *Lichtenau*, *Lichtenfels* and *New Hernet*.

BRITISH AMERICA

Includes New Britain, Canada East, and Canada West, New Brunswick, Nova Scotia, and Newfoundland.

New Britain is noted for its cold climate, for its savages or Esquimaux Indians, and for its being the ground or theater for the operations of the Hudson Bay Company.

The natives live mostly on seal. In traveling, they are drawn by a very fine kind of dog, which is harnessed to their sleds.

The British have trading stations at the mouths of most of the rivers, where the Indians come to exchange their furs for blankets, guns, beads, &c.

COUNTRIES AND TOWNS.

Canada West.

Kingston is found N. E. Ontario's roar,
While west the lake is Hamilton in Gore;
Toronto to the N. W. finds a Home,
As north the lake, Port Hope and Coburg come.

Canada East.

In the lower Province, Montreal lives ever,
Upon an island in *St. Lawrence River;*
And down the stream, one hundred eighty miles,
Quebec to heaven heaves up her giant piles;
A fortress strong, on a high promontory,
And famed in song, in history, and story.

New Brunswick.

St. Johns, and Frederickton, New Brunswick keeps,
Where the *St. Johns* along her pathway sweeps;
West of St. Johns, and east the young St. Croix,
St. Andrews lives, by commerce her employ.

Nova Scotia.

Yarmouth, on Nova Scotia's western border,
Where *Fundy's tide* rolls up in wild disorder,
And Halifax, near the peninsula's center,
Known for her port, where the largest crafts may enter;
With vessels of all kinds, this place is full,
But chiefly with the war ships of John Bull.
The northern shores, which Pictou has a seat on,
With Sidney, on the island of Cape Breton,
Are noted for their coal the world all over,
Which mineral freights full many an ocean rover.

Canada West.

Noted as being the most productive, and best settled of the British Possessions in N. A.; also for the intelligence of its inhabitants, who are mostly of an English origin.

The soil is a fine dark loam, mixed with a vegetable mould, and is unsurpassed for its productiveness.

Canada East.

The cultivated portions lie in the valley of the St. Lawrence.

It has the climate of Sweden, though the latitude of France.

The French language is chiefly spoken.

New Brunswick.

Noted for its immense quantity of lumber; mostly pine.

Frederickton on the St. Johns River, is the capital.

St. Johns is the largest town.

St. Andrews is the second in population.

Nova Scotia.

A peninsula south of New Brunswick, having a rugged stony coast. Noted for coal and gypsum. Climate mild; subject to fogs.

NEWFOUNDLAND.—A barren, hilly island in the Gulf of St. Lawrence, abounding with good harbors, and noted for the greatest codfisheries in the world.

An Iceberg.

UNITED STATES.

Noted as being the largest, most enlightened and powerful republic on the globe. Bounded by the Atlantic on the east, and the Pacific on the west. Having an area of more than 2,000,000 square miles.

I.—EASTERN STATES.

Maine.

In Maine, Augusta, on the *Ken-ne-beck*,
Just 50 miles, if right I recollect;
There Hallowell, for *granite* we 'll remember,
And Bath, for building *ships* of white oak timber.
Ban-gor' in *lumber* trades; as *boards* and *plank*,
And takes her place upon *Pe-nob-scot's* bank,
O-ro-no, Frankfort, Bucksport, and Castine (teen),
On the same banks, by the same glowing stream.
Portland by Casco Bay, chief town in Maine,
In fisheries and *commerce* holds her reign.

New Hampshire.

Portsmouth her harbor boasts, and sits supine,
Where meets *Pi-scat-a-qua* the rolling brine.
Up the same tide is found Great Falls and Dover;
As on *Connecticut* is fair Hanover.
And Concord on the *Mer-ri'-mack* may rest her,
As further south is Nash'-ua and Man-ches-ter.

Maine.

Noted for its vast forests of lumber, for ship building, and for being the most northeastern state in the union.

It was a part of Massachusetts till 1820, when it became a separate state.

The climate is cold and healthy.

The coast is rugged, and the harbors numerous.

A large part is still covered with forests.

It is better adapted to grazing than agriculture.

AUGUSTA is noted as the capital.

Hallowell for granite quarries.

Bath for ship building.

Portland for commerce and fisheries, and as being the largest town in Maine.

New Hampshire.

Called the Granite State. Noted for the White Mountains.

The surface is level on the coast.

It abounds in rivers, lakes and mountains.

The soil is better fitted for grazing than tillage.

The climate is cold and healthy.

CONCORD is noted as the capital

Vermont.

Montpelier, near the center of the state,
On *Onion River*, rules o'er small and great.
Windsor upon *Connecticut* may reign,
As Burlington is found on *Lake Champlain*,
And *Otter Creek* has fair Vergennes' upon her,
Known for the *fleet* of Commodore McDonough;
And Mid-dle-bu-ry on this tide may tarry,
Known for her *college* and her *marble quarry*.
And Bennington, southwest of all, we mark
Famed for the victory of General Stark.

Vermont.

Noted for the Green Mountains, and for the enterprise of its hardy inhabitants.

A large portion of it is still covered with timber.

The valleys are well adapted to tillage and the hills to grazing.

The manufactures are not extensive.

The climate is cold and the winters severe.

MONTPELIER is noted as the capital.

Vergennes, as being the place where McDonough fitted out his fleet for the celebrated battle on Lake Champlain.

Middlebury, for its college and fine marble quarries.

Massachusetts.

Boston and Charlestown both together lay,
With Cambridge, born by Massachusetts Bay.
Lynn, famed for shoes, for codfish Marblehead,
Salem for wealth, gained in the India trade.
Gloucester for *mackerel* and *codfishing* both,
And Newburyport, for *commerce* farthest north.
Lowell on Merrimack, a far famed weaver.
For which is known both Taunton and Fall River.
As Springfield, for her *armory*, we hail.
New Bedford and Nantucket fish for *whale*.
Plymouth, known for the Pilgrim Fathers' landing,
By Cape Cod Bay, in Massachusetts standing.
And Worcester, that 's near the Bay State's center,
As a great thoroughfare, we next will enter.

Massachusetts.

Called the Bay State. Noted for its wealth, and the active part it took in the great struggle for the liberty and independence of our country.

The eastern part is uneven, and the western mountainous. Mount Tom, Mount Holyoke, Saddle Mount and Wachusett, are celebrated peaks.

BOSTON is noted as the capital, and as the largest town in New England.

Charlestown for the Bunker Hill Monument.

Cambridge for its university.

Lynn for the manufacture of shoes.

Marblehead for its cod-fisheries.

Salem for its wealth, obtained in the India trade.

Gloucester for cod and mackerel-fisheries.

Newburyport, the most northern town in the State, for commerce.

Lowell, as the first town in manufacturing in the United States.

New Bedford and Nantucket for whale fisheries.

Plymouth for the landing of Pilgrim Fathers, 1620.

Connecticut.

And *Hartford*, Middletown, and Say'-brook bide,
Fast by *Connecticut's* unfailing tide.
New London, Bridgeport, Fairfield and *New Haven*,
With Norwalk by *Long Island Sound* are graven.
And Stonington, southeast of all, we hail,
That with New London, fish for *seal* and *whale*.

Connecticut.

Noted for the ingenious character of its inhabitants, and for its schools.

The common school fund, in this state, is over two millions of dollars.

It has been distinguished for its men of genius and learning.

HARTFORD and NEW HAVEN are the capitals.

Rhode Island.

Among her factories, Providence makes her stay,
On her own stream, by *Nar-ra-gan-sett Bay:*
And Newport, on Rhode Island finds resort,
Well fortified and noted for her *port*.
From Providence north, Pautucket finds a seat,
As north of Newport, Bristol next we greet.

Rhode Island.

Noted as the smallest state in the union.

It consists mostly of the shores and islands of Narragansett Bay, that gives it great advantages for navigation.

PROVIDENCE is noted as the capital, likewise for its manufactures, as well as being the seat of Brown University.

Newport, as a naval station, for its fortifications and its excellent harbor.

II.—MIDDLE STATES.

New York.

In New York State, where Hudson meets the brine,
New York and Brooklyn in their trade combine.
On the same tide, West Point and Newburg stay:
Poughkeepsie, Hudson, Troy, and Albany.
Schenectady, with Utica and Rome,
Upon the *Erie Channel* find a home.
Here Syracuse and Rochester, we see—
The last is on the River Genessee.
Then Brockport comes, with Lockport in the score;
As Buffalo is found on Erie's shore.
From Buffalo east, takes Attica her fare;
In Genessee, Batavia has a share.
Then Canandaigua in Ontario view;
As stands Geneva east, with Waterloo.
Auburn is seated by Owasco tide.
South of Cayuga, Ithaca is spied.
Oxford and Norwich in Chenango trace;
In Courtland County, Courtland has a place.
Bath in Steuben, Elmira in Chemung;
Owego next, then Binghampton in Broome.
Near Saratoga, Ballston makes her quarters,
And both are noted for their mineral waters.
Salem in Washington, with Sandy Hill;
Whitehall is where Champlain's dark waves distil.
Ticonderoga lives by Lake Champlain, [reign.
Where stands Crown Point, and Plattsburg holds her
Oswego sits beside Ontario's border;
While on the eastern coast is Sackett's Harbor.
A place to Watertown, Black River warrants;
As Ogdensburgh is found upon St. Lawrence.

New York.

Called the Empire State. Noted for its canals, railroads, extensive commerce, and its great political influence.

Its population is greater than any other state in the union.

The route from New York to Buffalo, is one of the greatest thoroughfares in the world.

The scenery on the Hudson is of a sublime and imposing character.

The steamboats on this river are celebrated for speed and grandeur.

ALBANY, on the Hudson, is noted as the capital.
New York, at the mouth of the Hudson, as being the largest, most commercial, and important town in America.
West Point for its military academy.
Sing Sing and Auburn for State prisons.
Utica for the State Lunatic Asylum.
Schenectady for Union College.
Syracuse for its salt works.
Rochester for its flouring mills.
Lockport for its costly and expensive canal locks.
Buffalo as one of the most commercial towns in the United States on the lines of two the greatest thoroughfares in America—the Erie Canal, and Niagara and Lake Ontario routes.
Ballston and Saratoga for mineral waters.
Oswego as the principal port on Lake Ontario.
Sacketts Harbor for a battle fought during the last war with Great Britain.
Watertown for its neatness and manufactures.
Ogdensburgh as lying adjacent to Canada East.

The city of New York is one of the most important towns in the world. It is situated at the mouth of the Hudson, on an island about fifteen miles in length.

It is visited by vessels from all parts of the world. Mail steamers are daily arriving from, or departing for foreign ports.

Broadway is one of the finest streets in the world and the Park Fountain one of the city's greatest ornaments. Among the public buildings we may mention Trinity Church, a gothic structure, having a spire or steeple, 283 feet high.

New Jersey.

Trenton, that takes from Jersey's shore her fare,
Is on the eastern side of *Delaware.*
Then Bordentown, from Trenton south is seen,
With Burlington and Camden down the stream.
Freehold in Monmouth, known for Monmouth battle.
Princeton N. E. from Trenton, deigns to settle.
Where River Raritan pours forth her waters,
New Brunswick stands, and Amboy makes her quarters.

New Jersey.

Noted for manufactures, canals, railroads and its revolutionary incidents.

The southern part is barren and thinly settled; the northern part, rough and mountainous; the middle part is extremely fertile.

Apples and peaches are raised in great abundance in the southern part.

The Philadelphia and New York markets are supplied with their best fruits from this State.

TRENTON, on the Delaware, is noted as the capital.
Freehold is noted for the Battle of Monmouth, fought 1778, between the British under Lord Cornwallis, and the Continental army under Gen. Washington.

From Amboy north, Elizabethtown we view;
In Morris, Troy and Morristown are two.
Upon Passaic's banks, Newark has grown;
As farther up the stream is Patterson.

Trenton and Princeton are likewise celebrated for battles fought during the Revolution, in all of which Washington commanded in person.

Newark, noted for the manufacture of shoes and carriages.

Patterson, noted for its cotton manufacturing.

Pennsylvania.

PITTSBURGH.

There *Schuylkill* and the *Delaware* convene,
Is Philadelphia, oldest child of Penn.
And Harrisburg, the state metropolis,
On Susquehannah River, none can miss,
'Bove Harrisburg, Northumberland may tarry:
As in Luzerne is one, called Wilkesbarre.
Mauch Chunk in Carbon, where the *Lehigh* pours;
Honesdale in Wayne, where *Lackawaxen* roars.
Milford, in Pike, and Stroudsburg in Monroe,
Easton where *Delaware* and *Lehigh* flow.
Bristol in Bucks, 'bove Philadelphia's landing;
Chester below, in Delaware is standing.
From Philadelphia west, *three score* and *two*,
Lancaster, in Lancaster county, view.
On Schuylkill banks, is one called Norristown;
There Reading keeps; there Pottsville sits her down.
From Reading west, is Lebanon the while;
York lives in York; in Cumberland Carlisle.
From Cumberland, is Chambersburg southwest;
As south in Adams, Gettysburg may rest.
Where Alleghany joins Monongahela,
Pittsburg is found, though smoke and coal conceal her;

Pennsylvania.

Noted for coal mines and iron manufactures; and as being the center of the Alleghany Mountains.

The mountains extend through the middle of the state, leaving the northeastern and southwestern portions level, or undulating.

It is the first manufacturing state in the union, and is the richest in minerals.

The iron mines of this state are great sources of wealth, and chiefly supply the manufacturing establishments.

The coal beds are inexhaustible, yielding over two millions of tons annually; and in real importance and worth, are more valuable than the gold mines of Mexico, or California. On the eastern side of the mountains is found the anthracite, or hard coal; on the west bituminous, or soft coal. Pittsburg is near the center of the bituminous coal region.

Wheat is the principal product of the soil, though corn and other grains are raised in great abundance.

Its population is second to none but New York.

Several battles were fought in this state during the revolutionary struggle. Valley Forge, twenty miles northwest of Philadelphia, is known as the place where Gen. Washington made his winter quarters, during the darkest hours of the contest.

HARRISBURG, on the Susquehannah, is the capital.

Philadelphia is noted as being the largest town in the state, and second in the United States. It is distinguished for its humane and literary institutions; among which are Girard College, for orphans, the School for the blind, and one for deaf and dumb persons. Fairmount Water Works, which supply a great portion of the city with pure water from the Schuylkill river, hold a prominent place among the pleasure grounds, which are numerous.

Pittsburg, the second town in the state, in population, is distinguished for coal mines in its vicinity, and for the manufacture of iron, glass, white lead and heavy machinery.

Pottsville, Mauch Chunk and Honesdale, are noted for their coal mines.

Reading is a large and beautiful town, situated about fifty miles from Philadelphia, on the Schuylkill river. It is distinguished for its extensive iron works.

Easton is noted for its flour mills.

Wilkesbarre for the massacre of the inhabitants in the Wyoming valley, during the Revolution.

The works of internal improvement in this state have greatly facilitated the intercourse with the eastern and western portions of the union.

Since 1844 the system of common school education has received its due attention.

Here Birmingham, and one called Alleghany,
Their stations take in Western Pennsylvania.
Erie is where Lake Erie's waves roll ever;
As Beaver lies upon the Ohio River.

Delaware.

On *Jersey's Creek*, in Delaware, is Dover,
While *Brandywine*, fair Wilmington lives over.
And Delaware City, with Newcastle fair,
On the west bank of River *Delaware*.

Delaware.

Noted as having the smallest population of any state in the union, and the smallest territory excepting Rhode Island.

In the northern part the soil is fertile; in the southern unproductive.

On the Brandywine there are extensive establishments for the manufacture of paper, gunpowder, cotton and woolen goods.

Dover is the capital.

Wilmington is noted as the largest town in the state, as well as for its great water power; which is used in propelling flour, paper, powder and cotton mills.

III.—SOUTHERN STATES.

Maryland.

Of Maryland's towns, the first is Baltimore
Near Chesapeake, upon *Pa-tap-sco's* shore.
And west from Baltimore, miles forty-two,
Has Fredericktown *Mo-noc'-a-cy*, in view.
On *Severn's* bank, two miles from Ches-a-peake,
An-nap'-o-lis rules the powerful and the weak.
A German settlement is Ha'-gerstown,
West of the Blue Ridge is her station known.
And Cumberland of *Potomac* may share,
'T is west of all I've named, a thoroughfare.
And Washington, three hundred miles from sea,
On the east bank of *Potomac* doth lay.
Georgetown from Washington, *Rock Creek* divides;
West of *Potomac* Alexandria hides.

Virginia.

The staple production of Virginia, east of the Blue Ridge, is Tobacco.

One hundred fifty, from the mouth of *James*,
In old Virginia, Richmond makes her claims.

Maryland.

Noted for mild climate, favorable situation for commerce, and as having been settled by Roman Catholics.

It is separated from Virginia by the Potomac, and divided into two parts by the Chesapeake Bay. These waters are navigable to the extreme boundaries of the state.

The soil is rich and produces an excellent quality of wheat.

Annapolis is the capital.

Baltimore is noted as being the largest town in the state, and the fourth in the union. It is the greatest flour market in the world.

The District of Columbia was ceded to the United States' government, by Maryland and Virginia, in 1790. It is ten miles square, containing an area of one hundred square miles.

Washington is noted as being the capital of the United States. It is situated on the east bank of the Potomac, which is navigable to this point for ships of the line. An United States' navy yard is also established here.

Virginia.

Noted as the birth-place of the immortal Washington, and for having given six presidents to the Union.

It is crossed by the Alleghany Mountains and Blue Ridge, which extend N. E. and S. W.

The soil, on the coast, is sandy and sterile; on the banks of rivers and in the valleys, it is rich and fertile.

The climate, on the coast, is unhealthy; but, among the mountains, cool and salubrious.

Richmond is the capital of Virginia.

Norfolk has a fine harbor, and noted for foreign commerce. On the opposite side of the Elizabeth river, is Gosport; noted for the United States' Navy Yard, and an extensive dry dock.

Yorktown is noted for the surrender of Lord Cornwallis to General Washington, in 1781.

From Hampton Roads, eight miles, Norfolk lives ever,
Upon *Elizabeth's* fair, flowing river.
Yorktown, upon *York river*, sits alone,
Where Lord Cornwallis bowed to Washington.

North Carolina.

Ra'-leigh, in the interior of N. C.,
Near *river Neuse*, whose waves are ever free.
Newbern, the largest town, stands by the *Neuse;*
Of *Cape Fear river*, Fay'-ette-ville makes use.
And Wilmington comes in the stanza next,
Up *Cape Fear river*, miles—just thirty-six.

South Carolina.

Columbia stands upon the *Con-ga-ree';*
And Georgetown dwells upon the *Great Pedee'*,—
As Charleston lives just seven miles from sea.
Hamburg, by the *Sa-van'-nah*, keeps her station,
Just at the head of steamboat navigation.

Georgia.

Where rolls *O-con'-ee's* waves, is Mil'-ledge-ville;
Augusta, where *Savannah's* waters spill;
On the same tide, Savannah mounts her bluff—
Say, seventeen from sea, for that's enough.

Alabama.

On *Alabama's* breast, Montgomery 's hung;
As Tus-ca-loo'-sa hears *Black Warrior's* song.
Mobile is west the mouth of *Mobile tide;*
As Blakely lives upon the other side.

Mississippi.

Jackson, in Mississippi, drinks the *Pearl;*
Tombigbee's waters round Columbus furl;
And Natchez makes a rising bluff her seat—
O'er *Mississippi's* wave, three hundred feet.
On the *same tide*, below the dark *Yazoo*,
From Jackson, west, Vicks-burgh is in the view.

Louisiana.

And New Or-leans' o'er Louisiana smiles,
Up *Mississippi's* stream, one hundred miles.
'Bove New Orleans one hundred, ten, at most, [110
Is Ba-ton Rouge, a military post.

North Carolina.

Noted for its Gold Mines, that yield $4,000,000 annually; and, also, for the Dismal Swamp, thirty miles long and ten miles wide.

It is low and sandy, for sixty or seventy miles from the coast. In the interior, it is hilly; and in the western part it is mountainous.

This State has no good harbors. The coast is lined with sand bars and reefs, that render navigation dangerous.

RALEIGH, near the center of the State, is the capital. Newbern is noted as the largest town in the State. Wilmington, the chief commercial depot, is noted as being the terminus of an extensive line of railroad.

South Carolina.

The smallest of the Southern States; noted for the opulence and independent character of its planters.

COLUMBIA, the capital, is noted for its neatness. Charleston is noted as the largest of the Atlantic cities in the Southern States.

Georgia.

Noted for its productions of rice and cotton, as well as for gold mines. The mines are found in the northern part.

In surface and soil, it resembles the Carolinas. Indigo was once raised in large quantities, but cotton has now taken the lead of all other products. Sugar cane is raised to some extent in the southern part.

MILLEDGEVILLE, the capital, has a pleasant situation. Savannah is noted as being the largest town in the State, and as having a large share of commerce.

Alabama.

Noted for its fertile soil, and rapid increase in population.

Cotton is the chief agricultural product.

MONTGOMERY, on the Alabama river, is the capital Mobile is noted for its extensive commerce. Tuscaloosa, the former capital, is a flourishing town

Mississippi.

Situated mostly in the basin of the Mississippi river, which bounds it on the west. It is noted as being the chief cotton growing state in the Union.

The southern portion is level, and the northern mountainous.

JACKSON, the capital, is on Pearl river.

Natchez, the largest town, and a place of great trade, is situated on a high bluff, 300 feet above the Mississippi.

Louisiana.

Noted for its great commercial advantages, and as the chief sugar growing state in the Union. It is a low and level tract, and at the southern part forms the delta of the great Mississippi. The waters of the river are higher than the surrounding country, and are kept from overflowing by levees.

And Alexandria, for its trade in *cotton*,
Upon *Red river*, must not be forgotten;
And Natch-i-toches, an old French town we enter,
On the *same tide*, near Louisiana's center.

The sugar raised in Louisiana, in 1845, was 207,000,000 of pounds.

NEW ORLEANS, by far the most important town in the Southern States, is situated on the Mississipi, one hundred miles from its mouth. Its commerce is next to that of New York.

Baton Rouge, the present capital, is noted as a military station, and for a United States arsenal.

Alexandria is noted for its trade in cotton.

Florida.

Scene in Florida.

In Florida is Tal-la-has'-see found,
The seat of rule, on elevated ground,
And Pen-sa-co'-la stands, far to the west,
And of the *Gulf* ports she is deemed the best.
Upon the eastern coast, St. Augustine [*au-gus-teen'*
Oldest of towns, beneath a sky serene.

FLORIDA is the native state of the Seminole Indians. It is noted for fertility of soil and luxuriant vegetation, and as being the most southern part of the United States. It formerly belonged to Spain, but was ceded to the United States in 1819.

TALLAHASSEE, the capital, has an elevated site.

St. Augustine is noted as the oldest town in the United States. It is the resort of invalids, on account of its mild and healthy climate.

Pensacola is noted as a man-of-war station, and for its harbor—the best in the Gulf of Mexico.

Florida.

OCEOLA, SEMINOLE CHIEF.

The above cut is a representation of OCEOLA, the Chief of the Seminole Indians, who long maintained a bloody and even-handed war with the United States. They inhabited the Everglades of Florida, and were assailed in vain, till hunted down by blood hounds procured from Cuba.

Texas.

Austin, from *Colorado*, slakes her thirst,
And o'er the Lone Star reigns supreme and first.
Bas-trop', La Grange, Columbus further south,
With Mat-a-gor-da at the river's mouth.
Sa-bine' is at the mouth of dark *Sa-bine*, [*sa-been*
And Gal'-ves-ton beside her bay is seen.
Houston, northwest of Galveston, we view,
On a small stream, call'd *Buffalo Bayou.*
Ve-las'-ca, where the *Brazos* meets the brine—
A place of much resort in summer time.
Go'-liad, a town on *San An-to'-nio's* snore,
Where Fannin died, with full four hundred more.

Texas.

The Lone Star—noted for its rapid increase in population, and for having once been a part of Mexico.

It was declared an independent state in 1835; and, in 1846, it was annexed to the United States.

Texas contains about six times as much territory as Pennsylvania.

The chief productions are cotton, tobacco, rice, sugar, corn and wheat.

It abounds in buffalo, deer, and wild horses, that roam over its vast plains.

The inhabitants are mostly from the U. States.

AUSTIN is the capital.

Sabine City, on the river Sabine, is a port of entry.

Houston is the most commercial town in the State.

Velasco is noted as a summer resort.

Goliad is noted for the massacre of Col. Fannin, and four hundred prisoners, by the Mexicans.

IV.—WESTERN STATES.

Arkansas.

Twice thirty miles, up the *Arkansas'* billow,
O'er the north bank, Arkansas makes her pillow.
Arkansas State to Little Rock lays claim,
That by *Arkansas* river writes her name.
Van Buren, near the Indian Territory,
Up the *same tide*, is woven in the story.
And Bates'-ville slumbers on the river *White*,
As, in the northwest, Fay'-ette-ville we write.

Arkansas.

Lies west of the Mississippi, and noted for hot springs.

It was admitted into the Union in 1836.

Along the Mississippi, it is low and unhealthy. The interior is elevated, and enjoys a salubrious climate.

It abounds with large rivers.

LITTLE ROCK, the capital, is on Arkansas river. Arkansas is noted as being the oldest town in the state; settled by the French in 1685. Its population is mostly descendants of French and Indians.

Tennessee.

Nashville, of TENNESSEE, is known to stand
By the Great Bend of *River Cumberland*.
As Gal-la-tin near *Cumberland* is seen,
Clarksville, on *Cumberland*, is down the stream;
As South from Nashville, Franklin takes her throne;
From Nashville East is stationed Lebanon.
And where *Duck River* rolls her purling rill,
Columbia stands with one called Shel-by-ville.
In RUTH-ER-FORD, as Mur-freys-bo-ro smiles;
Near Fay-ette-ville, Pu-las-ki lives in GILES.
Kingston in ROANE; and Athens in McMINN;
Knoxville in KNOX; as Greenville lives in GREEN.
Memphis, secure from *Mississippi's* waters,
South West of all makes a *high bluff* her quarters.
And Pur-dy, Bol-i-var, and Ra-leigh, three,
Stand with La Grange in WESTERN TENNESSEE;
Where Somerville may learn her pedigree.
And North of these Brownsville and Jackson trace,
Where Trenton comes, and Pa-ris finds a place.

Noted for the Cumberland Mountains; for its healthy climate and fertile soil. It is divided, by the Cumberland Mountains, into East and West Tennessee.

Kentucky.

Where rolls the *Ohio*, Mays-ville let us greet;
In BRACK-EN, there Au-gus-ta finds a seat.
New-port and Cov-ing-ton are side by side,
Where *Lick-ing River* joins her parent tide.
War-saw is where the Ohio billows range
With Car-roll-ton, one Bedford and La Grange:
Here Lou-is-ville, the largest town, is seen
With Bran-den-burg, that's further down the stream;
And Haws-ville, here in HANCOCK County, ranks;
As Ow-en-bo-ro mounts the *Yellow Banks*:
Then Hen-der-son and Smith-land, each, are passed;
As comes Pa-du-cah in the stanza last.
Frankfort is by Kentucky's purling rill;
In SHELBY West, is one called Shel-by-ville;
Georgetown in SCOTT; in NICHOLAS Car-lisle;
As HARRISON sees Cyn-thi-a-na smile.
Paris, in BOUR-BON, makes her quarters yet,
While Lexington is stationed in FAY-ETTE.
And Nich-o-las-ville, in JES-SA-MINE, we mark;
Versailles in WOODFORD; Win-ches-ter in CLARK;
Rich-mond in MADISON; while to GER-RARD,
The town called Lan-cas-ter, we next award.
Stan-ford in LIN-COLN; Danville then in BOYLE;
As Har-rods-burg in MERCER lives the while.
Then Taylorsville and Shepardsville we greet,
While Springfield makes in WASHINGTON her seat.
In NELSON Bardstown; Greensburg lives in GREENE;
As Ma-ri-on stands with Leb-a-non, between.
Columbia in A-DAIR; Glas-cow in BAR-REN;
As Bowling Green her station makes in WARREN.
And Rus-sel-ville, Elk-ton and many more,
With Hopkinsville and Princeton join the score.

Settled by Daniel Boone, in 1769. It is noted for its delightful climate and fine soil, the Mammoth Cave, and for the brave and hospitable character of its inhabitants.

Indiana.

Of Indiana State, the ruling Miss,
Upon *West Fork*, is In'-di-an-ap'-o-lis.
Known for her vineyards, by the *Ohio's* tide,
Where lives New Albany, is Vevay spied.

Indiana.

Smallest of the Western States, though one of the most fertile and prosperous.

It resembles Ohio in surface, soil and climate.

The people of this state, as well as in all of the Western States, are chiefly employed in agricultural pursuits.

On the *same tide*, makes Madison her lair ;
Where, from Columbus, drives the railroad car.
With Lawrenceburg let Cambridge City mix ;
Though space divides them, miles, just seventy six.

Vincennes, Terre Haute, and Covington, we rank [*tere hote*]
With La Fayette, all on the *Wabash* bank.
There, Delphi keeps ; there, Logansport is known ;
There stand Peru, Wabash, and Huntington.
Northeast from this, bound by the tow-path chain,
Where foams *Maumee*, in Allen, is Fort Wayne.
South Bend is where *St. Joseph's* billows break ;
As Michigan City smiles above the *Lake*.

Indian corn, wheat, oats, beef and pork, are raised in vast quantities, with half the work necessary in the Eastern States.

INDIANAPOLIS, on West Fork, of White river, is the capital.

Vevay is noted for its vineyards, planted by Swiss emigrants.

New Albany is noted as being the largest town in the state.

Michigan City is the only good port in the state, on Lake Michigan.

Ohio.

Columbus reigns upon Ohio's soil,
While at her feet *Scioto's* waters boil.
With Steubenville, and one called Marietta,
On the *Ohio tide*, is Cincinnati.
And Sidney, Troy, and Dayton, find a home,
Upon *Miami's* banks, with Hamilton.
Cleveland is north, where *Erie* chants her ditty ;
As, west from Cleveland, stands Sandusky City.
Lower Sandusky, where *Sandusky's* billow
Gives Tiffin, and Bucyrus, each, a pillow.
In Ot-ta-wa', Port Clinton, finds a spot ;
Huron, in Erie, must not be forgot.

Norwalk, and Mansfield, with Mount Vernon, lain
From Erie, south ; while Wooster lives in Wayne.
And south of Wayne is Millersburg the while ;
As, north, Medina, and Elyria, smile.
Northwest of all, where foams the dark *Maumee*,
Charloe, Defiance, and Napoleon see :

And Perrysburg lives by the *Maumee stream* ;
Where, further down, Toledo's turrets beam.

With Bryan, Paulding and Van Wert, are west,
Celina, Greenville, Eaton, and the rest.
Putnam and Findlay, near the dark *Auglaize* ;
Lima, and Kenton, both, are south of these.

Dresden and Zanesville, o'er *Muskingum* bide ;
McConnellsville is seated down the tide.

Logan, and Athens, on the *Hocking* stay ;
Jackson, and Chester, south of them may lay.
With Pomeroy, Galliopolis lives ever,
By Burlington, on the *Ohio river*.

From Cleveland, south, along the tow-path side,
Cuyahoga Falls with Akron are espied ;
Fulton and Massillon we likewise view ;
Then Bolivar and Philadelphia too :
Coshooton next, then Newark, Circleville,—
The last is where *Scioto's* waves distill ;

Ohio.

Called the Buck Eye State. It is destitute of mountains, though hilly and irregular along the valley of the Ohio river.

Large prairies are found at the head waters of the Scioto and Muskingum.

This state is noted for great wealth and rapid increase in population.

Lake Erie and the Ohio river give it great advantages for commerce.

It was settled as late as 1789, yet, in point of population, is the third state in the Union.

COLUMBUS, the capital, is situated on the east bank of the Scioto river. Its site was selected in 1812, as the seat of the legislature, and was then an entire wilderness.

Cincinnati, situated on the Ohio river, in the southwest part of the state, is one of the largest cities west of the Alleghanies. It is one hundred and fifteen miles southwest of Columbus, four hundred and ninety from Washington, and about nine hundred from the city of New York, by the Buffalo and Lake Erie route. This city, in 1795, contained but 500 inhabitants; in 1800, 750 ; in 1820, the population was 9,640 ; in 1830, 24,000 ; in 1840, 46,000 ; in 1847, the population had reached as high as 90,000; and, at the present time, probably exceeds 100,000.

The climate of this city is subject to considerable extremes of heat and cold, but is generally considered healthy.

Cincinnati is noted and distinguished as being the greatest pork market in the world.

The streets that run east and west are denominated First, Second, Third, Fourth, &c. ; those that run north and south are named ; as, Elm, Race, Vine, Walnut, Main, &c. The city is between the river, on the south, and a high hill surrounding it on the east, north and west ; the streets that run north and south extend from the river to the hill.

Cleveland is the principal port for this state, on Lake Erie. Its advantages for trade and commerce are great. The city, with the exception of that part bordering on the Cuyahoga river, is one of the most beautiful in the United States. The streets are all wide, and the houses are neat and beautifully shaded with trees.

And Chillicothe standing further south,
Drinks from this tide with Piketon and Portsmouth.

East from *Miami*, west *Scioto's* furrow,
Are London, Washington, and one Hillsboro.
In Clinton, Wilmington; (West Union count,)
Georgetown in Brown, Batavia in Clermont.
Urbana lies, with Springfield by her side,
Where roars *Mad river*, in its might and pride.
Xenia, from Springfield, south, is found in Greene; (*Zenia*)
In Warren county, Lebanon is seen.
Bellefonte, from Springfield north, in Logan know;
As Woodfield, east, is stationed in Monroe.
And Marysville, Marion, and Delaware,
Near the glad waters of Scioto fare.
Lan-cas-ter lives in Fairfield county yet;
While, in the county east, is Somerset.
In Guernsey county, Cambridge numbers one;
St. Clairsville next, Cadiz and Carrollton.
As, one New Lisbon rules Columbiana,
And Canton Stark; o'er Portage is Ravenna.
From Portage, north, Chardon and Painesville tread.
Warren is east, in Trumbull county bred.
Northeast of all, is Ashtabula known,
Whose county town is christened Jefferson.

Michigan.

Southeast the State of Michigan, in sight
Of her dark waters, is the town Detroit.
Adrian, Tecumseh and the fair Monroe,
Where *River Raisin* murmurs in its flow.
And Ypsilanti on the *rail-way* keeps
In Washtenaw, where *Huron River* sweeps.
Ann Arbor then with Dexter we may view,
Then Jackson comes, Marshall and Kalamazoo.
Paw Paw is next, and last of all St. Joe,
Where the dark waters of *St. Josephs* flow.

And Hills-dale, Branch, and Niles, and Cen-tre-ville,
With Ber-ri-en, where *St. Joseph's* waters spill.
As Shelby, northward from Detroit, we track;
From Shelby, west, is seated Pontiac.
Near *St. Clair Lake*, Mt. Clemens seeks repose;
St. Clair is where the *St. Clair river* flows.
From Pontiac, west, Howell the first we scan;
Then Bellvue comes, Hastings, and Allegan.
Grand Haven, at the mouth of *river Grand*,
Just opposite Milwaukee, takes her stand.
Grand Rapids, with I-o-ni-a, up this stream,
Where Lansing lives, as capital, I ween.
Corunna, Flint, Port Huron, and La-Peer,
Are in the counties stationed east of here.
And Mackinaw keeps in an open *Strait*,
'Tween *Michigan* and *Huron*, 'tis the gate;
Upon a dusky *isle* her bulwarks flame,
A fortress strong, and owned by Uncle Sam.

Zanesville, on the Muskingum, opposite the mouth of Licking river, is a flourishing town.

Sandusky City is on Sandusky Bay, upwards of one hundred miles from Columbus.

Dayton, on the Miami, southwest of Wolf run, is considered one of the handsomest towns in the state. It is crossed by the Miami Canal that connects it with Cincinnati.

Chillicothe, on the west bank of the Scioto, has a beautiful situation.

Steubenville is in Jefferson county, in the eastern part of the state, on the Ohio river, and in a coal district.

The internal improvements in this state are rapidly progressing, and in extent are second to no state but New York.

The most important are as follows:

	LENGTH.
Ohio Canal and branches, ...	335 miles.
Miami Canal and branches, ...	84
Miami Extension Canal and branches,	138
Wabash and Erie Canal, ...	91
Walhonding Canal, ...	25
Hocking Canal, ...	56
Muskingum Improvement, ...	91
Mad River and Lake Erie Railroad, .	160
Little Miami Railroad, ...	140

Michigan.

Noted for its great commercial advantages, its mines of copper, forests of pine, and for its rapid improvement.

It consists of two great peninsulas, one between lakes Michigan and Huron; the other between lakes Michigan and Superior.

The copper mines are on the shores of Lake Superior.

Lansing, the new capital, is on Grand river, near the center of the state.

Detroit, the largest and most important town in the state, is favorably situated for commerce and trade, in the eastern part of the state, on the Detroit river. It is the half way house for boats and vessels running between Buffalo and Chicago.

Adrian, Tecumseh and Monroe, are flourishing towns on the Raisin river.

Ypsilanti is in Washtinaw county, on Huron river.

Ann Arbor, Dexter, Jackson, Marshall, Kalamazoo, Paw Paw, &c., are the principal places on the Railroad that crosses the state east and west. St. Joseph is at the mouth of the St. Joseph's river. Hillsdale, Niles' Branch, Centerville and Berrien are all in the southwest part of the state, on the same river.

Shelby is situated north of Detroit.

Grand Haven, at the mouth of Grand river, is opposite Milwaukee, in Wisconsin, on the western shores of the lake.

Mackinaw is noted for its fortifications, and for the annual meeting of the Indians, to receive their yearly stipend from the United States' government.

Illinois.

Springfield is capital of Illinois, (Illinoy)
Where river San'-ga-mon her notes employ.
Chi-ca'-go reigns the chief of all the clan,
With Little Fort beside *Lake Michigan.*
Lockport and Ju-li-et' with Dresden, twain,
Are near Chicago, on the dark *Des Plains.* (De Plain)

And *Illinois*, to Ot-ta-wa may roar,
Peru, Pe-o-ri-a, and Ha-van-na—four:
Then Beardstown comes, and Mer-e-do-sia's seen;
As, east this tide, is Carrolton, in Greene.

Ga-le'-na, noted for her mines of lead,
Northwest of all, by Fever River bred,
Rock Island, first on *Mississippi* view,
And then the Mormon city, called Nau-voo.
Then Warsaw comes, and Quincy next we rhyme;
And Al'-ton, noted for her *coal* and *lime.*
Kas-kas'-kia, a French town further south,
With Cai-ro seated by *Ohio's* mouth.

And Shelbyville, Vandalia, and Carlisle,
Along the banks of fair Kaskaskia smile.
Monmouth and Knoxville near each other rest;
Macon and Carthage, from Peoria, west.
And Rushville, Woodville, and Columbus, throng;
Near Quincy, is the place they all belong.
From Springfield, west, has Jacksonville her fare;
Known for the college that is stationed there.
And, by the *Wabash*, Danville sits her down;
While, on the *Ohio tide*, is Shawneetown

Missouri.

And Jefferson City on a high bluff smiles,
Up the *Missouri* tide twice sixty miles, (120)
On the same tide, just twenty from its mouth,
St. Charles is on the north bank, not the south,
And Independence, west of all hath laid her,
From whence for Santa Fé, leaves many a trader.
St. Louis, on the *Mississippi* 's seen,
Down from Missouri's mouth miles seventeen,
From New Orleans, twelve hundred up the tide,
Missouri's largest town, Missouri's pride.
And from St. Louis, seventy miles southwest,
Po-to'-si*lives, known for her *lead* the best.

Illinois.

Prairie on Fire.

The prairies of the Western States are every year swept over by fire. The view, when the tall grass is thoroughly dried and the flames are aided by a strong wind, is one truly magnificent and sublime.

Noted for rapid increase in population, and great fertility.

Agriculture is the chief employment of the people.

Lead is found at Galena, in the N. W. part of the state, in great abundance.

The canal, connecting Lake Michigan with the Illinois river, is now complete.

SPRINGFIELD, the capital, is on the Sangamon river.

Chicago, one of the largest towns in the West, is at the head of Lake Michigan.

Nauvoo, on the Mississippi, is noted as the Mormon City.

Alton is noted for its coal and lime.

Galena for its lead mines.

Missouri.

The largest state in the union, with the exception of Texas, and noted for its great mineral resources.

This state lies west of the Mississippi, and is intersected from west to east by the Missouri river, the great tributary of the Mississippi.

The mines of lead, iron, salt, coal, &c., are inexhaustible, and constitute the wealth of the state.

Iron Mountain is a mass of pure iron, 350 feet high, and two miles in circuit. Pilot Knob is another mass 600 feet high, and three miles in circuit.

JEFFERSON CITY, the capital, is on a high bluff, one hundred and twenty miles up the Missouri river.

St. Louis is one of the largest towns in the Western States, and bids fair to become one of the first in the union.

* Pronounced in English *Po-to'-si*; in Spanish, *Po-to-si'*.

Iowa.

Sac Indians Spearing Fish

The Sac Indians in this state subsist by hunting, trapping and fishing. The above cut represents them spearing fish

I'owa City sits the first in state,
Up her fair stream a cypher and an eight, (80)
From the state limits forty miles or more,
Is Burlington, on *Mississippi's* shore;
And north of this is Bloomington espied,
With Davenport upon the western side;
Ca-man'-che next, then Bell'-vue and Dubuqe',
Known for her *lead*, beside this giant brook.

Wisconsin.

Between two lakes holds Madison her rule,
And of the Badger State is capital.
Beloit and Janesville on *Rock River* bide,
As Prai-rie-du-Chien drinks *Mississippi's* tide;
And Mineral Point is near Potosi bred;
These two are noted for their mines of lead.
Lake Michigan, She-boy'-a-gan gazes o'er,
Milwaukee next, Ra-cine' and Southport four.
As Wau-ke-sha' we from Milwaukee track,
On Winnebago Lake is Fond du Lac',
And North of all, where the *Fox River* sweeps,
Upon Green Bay, Green Bay her station keeps.

Oregon.

This territory lies north of California, and between the Rocky Mountains and the Pacific Ocean. It is noted for being the great Western division of the United States; as well as for the enormous growth of its pines, which are sometimes found 250 feet high.

The soil, west of the Cascade Range, is represented as extremely productive.

Oregon City stands in a fertile valley near the falls of Willamette river; it contains upwards of 500 inhabitants.

Astoria is near the mouth of Columbia river.

Iowa.

The Northwest State of the union. Noted for its fertility and lead mines.

It is bounded on the east by the Mississippi river, which separates it from the states of Illinois and Wisconsin.

The soil is uncommonly fertile; large crops of corn, oats, wheat, &c., are raised with but little labor.

The lead mines of this state, with those of Wisconsin, Illinois and Missouri, are the richest in the world.

Iowa City, the capital, is on Iowa river.

Burlington is noted as being favorably situated for trade.

Dubuque is in one of the greatest lead districts in the world.

Wisconsin.

The Badger State. Bounded on the east by Lake Michigan, on the west by the Mississippi river. These waters give it great facilities for commerce.

It is noted for its valuable lead mines, its fertile soil, beautiful oak openings and numerous fine prairies.

The southern part of the state presents one of the best farming districts in the union.

The population is a multifarious mass of Europeans and Americans. The former are characterised for their industry and temperate habits; the latter for superior intelligence and enterprise.

Madison, between Third and Fourth lakes, is the capital.

Milwaukee, the largest town in the state, is noted for its rapid advancements in wealth, population and importance.

California.

Gold Digging in California.

This country was once claimed by Mexico, but was ceded to the United States by treaty, in 1848. It lies between the Rocky Mountains on the east, and the Pacific Ocean on the west.

It is noted for the vast quantity of gold found within its borders. The gold is dug from the mountains and rocks, and from the sand in the beds of the rivers.

MEXICO AND GAUTAMALA.

Mexico.

The City of Mexico.

MEXICO is one of the oldest cities on the Western Continent. It is situated seven thousand feet above the level of the sea, on the table lands. Upon the South are seen some of the loftiest peaks of the Cordilleras, and among them Popocatapetl, a large volcano that is crowned with perpetual snow.

And Mexico, high on the table lands,
In the interior of the province stands,
Above the sea full seven thousand feet,
Adorned with temples rich and structures great.
Fair lakes are there, arrayed in evergreen;
High mountain peaks upon the south are seen:
There Popocatapetl smokes all below,
From its high summit, covered o'er with snow.

The City of Vera Cruz.

Vera Cruz is noted for its Castle of San Juan D'Ulloa, one of strongest in the world, and which cost $40,000,000 to build it. It was taken by the United States' army under Gen. Scott, during the late Mexican war, but was restored by treaty.

For her castle famed, from Mexico due east,
Is Ve'-ra Cruz, three hundred miles, at least.
Southeast from Mexico, full eighty miles,
Famed for her churches, La Pu-e-bla smiles.
*Oa-xa'-ca, on this course, two hundred, stands,
Inhabited by numerous Indian bands;
While south, one, eighty, Ac-a-pul'-co keeps, [180
For her harbor known by the Pacific deeps.
In the interior, Gua-na-xua'-to shines, [*gwa-na-wha-ta*
With Zac-a-te'-cas near the silver mines.

*Wa-ha'-ca.

Mexico.

Noted for its mines of silver and gold, the former of which, have furnished more than half the silver of the known world.

It was once the seat of a powerful empire, over which presided a race of kings, termed "the Montezumas."

It was invaded and conquered by Cortez, a Spanish adventurer, in 1521; and became a province of Spain till 1821, when it was declared independent, and a republican form of government established.

Santa Anna.

The above cut represents the renowned Mexican leader in the war with the United States, and former president of the Republic.

Camanche Indian on Horseback.

The Camanche Indians, inhabiting the northern part of Mexico, are of a brave and hostile character. Mounted on their swift horses, they roam over the vast plains, attacking caravans and every thing that will afford them booty and plunder.

The Pyramid of Cholula.

The Pyramid of Cholula, in magnitude, rivals the great Egyptian Pyramid of Gizeh. It is only 250 feet high, but its length and breadth are 1335 feet, while the Egyptian Pyramid is only 728 feet. On its top was a temple dedicated to the sun. It was built of unburnt brick.

Que-re′-ta-ro, for beauty, has renown; [*ka-ra-ta-ro*
As, for her pyramid, Cho-lu′-la's known. [*co-lu′-la*
And Mat-a-mo-ras, on the Ri-o Grande, [*re-o-grand*
Just o'er the stream from Texas, takes her stand.
Re-sa-ca de-la Palm′-a's bloody ground,
With Pa′-lo Al′-to, north of this is found.
While west from here, twice eighty miles away, [160
On San Fernando's bank, is Monterey. [*mon-te-ra*
From Monterey, southwest, behold Saltillo,
Near Bue′-na Vis′ta's battle field her pillow.
From Mat-a-mo′-ras, south, Tampico smiles,
Along the coast two hundred eighty miles.
While west from here, San Louis Potosi, [*lue potosee′*
Upon *Tam-pi′-co* river makes her stay. [*tam-pee-co*

La Puebla is famous for its beautiful churches.

Oaxaca, 200 miles S. E. of the city of Mexico, is inhabited by bands and tribes of Indians.

Acapulco is noted for its beautiful harbor, which is the most capacious in the world.

Guanaxaco and Zacatecas are in the vicinity of the silver mines.

Queretaro is renowned as one of the most beautiful towns in Mexico.

Cholula is noted for its wonderful pyramid, made of unburnt brick.

Resaca de la Palma, Palto Alto, Monterey, and Buena Vista, are noted for victories gained by the United States' army under Gen. Taylor.

The gold mines of Mexico are inferior to those of Brazil, Peru and Siberia; but the richness of its silver mines is without a parallel, and have produced more of the silver coin than all the rest of the world put together. They are mostly found in the vicinities of Guanaxuato and Zacatecas.

The Mexican Mint is capable of stamping from 75,000 to $100,000 in one hour's time. It is estimated by some that the whole amount of money coined in this establishment exceeds three billions of dollars.

Lower California is a narrow peninsula in the northwestern part of Mexico. It is bounded on the north by Upper California, on the east by the Gulf of California, on the south and west by the Pacific Ocean. The soil is of a sandy, dry nature, and the population small.

Guatamala.

San Salvador, near the Pacific coast,
For indigo her trade is noted most;
Old Guatamala, once a splendid city,
Though of earthquakes now they sing a mournful ditty.
New Guatamala, five-and-twenty miles [25
From the old town, in wealth and commerce smiles.

Guatamala is a land of volcanoes; upwards of twenty volcanic peaks, in constant activity are seen in that part of the Cordilleras chain which is circumscribed within the territory of this province. The eruption of the Volcano of Casiguina, in 1834, was one of the most terrible and sublime upon record. The noise was heard for more than a thousand miles, and its ashes were carried more than eight hundred.

PRONUNCIATION.

Pensacola,	*Pen-sa-coo′-la.*	St. Augustine,	*St. Augusten′.*
Mobile,	*Mobeel.*	Natchitoches,	*Nash-i-tosh.*
Sabine.	*Sa-been′.*	Terre Haute,	*Tare Hote.*
Vera Cruz,	*Vera Cruse.*	Oaxaca,	*Wa-ha′-ca.*
Tampico,	*Tampe′-co.*	Monterey,	*Mon-te-ray.*
Resaca de la Palma,	*Re-sack′-a de-la Palm-a.*	Chihuahua,	*She-wau′-wau.*

Guatamala.

Noted for numerous volcanoes and frequent earthquakes, and for remarkable ruins found within its borders.

It was conquered by Spain in 1524, and held as a Spanish province until 1821; when it declared itself independent, adopting a republican form of government.

San Salvador, the capital, is situated in a very fertile valley, and is noted for its extensive trade in indigo and tobacco.

Old Guatamala was destroyed by an earthquake in 1775. It has sustained several shocks from the Water Volcano.

Balize Settlement.—This settlement was established and is owned by the British, for the purpose of cutting logwood and mahogany.

It extends along the Bay of Honduras 150 miles. It is inhabited by negroes, Indians and a few whites.

Balize, the capital, is a small town, which exports logwood and mahogany to the amount of $1,500,000 annually.

SOUTH AMERICA.

Commerce of the Andes, carried by Mules and Lamas.

Noted as having loftier ranges of mountains, larger rivers, a greater number of volcanoes, more extensive plains, richer mines of precious minerals, and sublimer natural scenery, than any other division of the globe.

Capes.

Cape Ve-la, first, by Mar-a-cay'-bo stands,
Most northern point of New Grenada's lands;
Orange and *North*, in *fifty* longitude,
Above the line, near French Guiana brood.
Five south, *five* west of *thirty*, *Cape St. Roque*, [*St. Roke*
Eastward of all, in Am-a-zo'-ni-a cloke.
'Tween twenty and the *line* called *Capricorn*,
St. Thomas and *Cape Fri'-ar* both are born.
From thirty-five to forty, as we go,
Are *Corientes* and *St. An-to'-ni-o*.
Cape Horn, near fifty-six, stands by his post,
On Ter-ra-del-Fu-e'-go's southern coast.
And *St. Fran-cis'-co* and *Cape Blanco* stay,
On the western coast of south America.

Rivers.

The *Port Desire* and *Cam-a-ra'-nes* fall,
With *Ri'-o Ne'-gro*, in the Atlantic hall.
The *Col-o-ra'-do* rolls her purpling billow,
From Buenos Ayres, with the dark *Saladillo*.
From here, *Salado* and *Vermejo* throng,
Where *Paraguay* drowns *Pilcomayo's* song.

PARAGUAY AND BRANCHES.

East of Bolivia, west of Par-a-guay',
In Am-a-zo'-nia born, pound to the sea,
O'er Buen-os Ayres. her rich and native home,
The *Paraguay'* and her dark branches come.
Round Paraguay, both south and east descried,
Rolls *Pa-ra-na'*, and empties in her tide;
Springing from Am-a-zo-ni-a's province wide.
'Tween Buenos Ayres and fertile Uraguay,
Named from the last, a river makes her way;
And *U'ra-guay* and *Paraguay*, are found
In *Ri-o de la Pla'-ta's* channel bound.

EASTERN COAST OF BRAZIL.

And eastward from the Amazonian coast,
In the same deeps, the *Diamond's* waves are lost.
The *St. Francisco* and *Salgado* there
With one *Par-na'-tha*, in his deeps appear.

AMAZON AND BRANCHES—SOUTHERN.

To-can'-tins, walled with *Ar-a-guay'*, rolls on
In *Pa-ra's* tide, or mouth of Amazon.
Xin'-gu, *To-pa'-jos* and *Ma-dei-ra* tread, (*zin-gu*)
With *Pu'-ros*, in the *Amazonian* bed.
Be-ni', *Ma-mo'-re*, *Blan'-co*, and dark *Gua-pore*,
Join in *Madeira* from Bolivia's shore.
And *Tef-fe*, *Ju'-rua*, and the *Ju'-tay* run, (*taf''-fa*)
From Sol-y-mas' to mother Amazon.
And from Peru, northward the *Ucayale*,
With dark *Hual-la'-go*, the same waters hail. (*wal-la'-go*)

NORTHERN BRANCHES.

From Eq-ua-dor', *Pa-tas'-co* southward goes,
And *Na'-po* there with *Pu-tu-may'-o* flows:
Pu-pu'-ra too, that skirts her limits north,
With *Ne'-gro* from Brazil here marshal forth.
While from Peru, and south of Eq-ua-dor',
Queen *Amazon* treads Amazonia o'er.

RIVERS NORTH OF AMAZON.

Ma-ro'-ni, *Su-ri-nam'*, and *Dem-er-ra-ra*, (*ma-roo'-ni*)
With *Es-se-qui-bo*, from Gui-an-a hurry.
From Ven-e-zue'-la, *O-ro-no-co's* rolled;
A-pu'-re there, a northern branch, behold:
Two branches more, we from Gren-a'da gather.
Me'-ta is one, *Gua-va'-ri* is the other.
And from Grenada, where Ca-rib'-bee storms,
The *Mag'-da-le'-na* rolls with *Cau-ca* in her arms.

TOWNS AND COUNTRIES.

Brazil.

Rio Jan-ei′-ro, in a country fair,
The capital, that breathes Brazilian air.
And Per-nam-bu′-co lives upon that shore,
With Mar-an-ham′ and fair St. Salvador.
And Rio Grande, so famous for its hides,
Where over *Palos Lake* the trader glides.

Diamond Washing in Brazil.

The diamond mines of Brazil are the most important in the world. They are worked by the government. The cut represents the manner in which they are washed from the sand.

Tejuco, for its diamonds bought and sold;
And Villa Ri′-ca, for her mines of Gold.

Guiana.

Georgetown, the English capital, may tarry
Along the mouth of *river Dem-er-ra′-ra.*
And Par-a-mar′-a-bo as Dutch we name,
Full eighteen miles up the hot *Surinam.*
Cayenne, well fortified, is east of all,
Upon an island, and belongs to Gaul.

Buenos Ayres.

And Buenos Ayres is on *La Plata* found,
The chief emporium of the province round;
San Ju′-an with Men-do′-za let us class,
Because each stands upon a *mountain pass.*
Cor-do′-va is an active trading town,
And Sal′-ta, for her *mules,* has much renown.

Paraguay and Uruguay.

San Car-los and Conception make their stay,
With fair As-sump′-tion, on the *Par-a-guay.*
In Uruguay, was Mon-ti-vi′-de-o born,
On the *La Plata* are her garments worn.

Brazil.

Noted as being the largest of the South American provinces; for rich gold and diamond mines, and for having been once governed by an European sovereign in person.

Its area is recorded as high as three millions of square miles.

Brazil was a colony of Portugal till 1812, when it was declared an independent state, and Pedro the son of the king of Portugal was, by the people of Brazil, made Emperor. The population is about five millions; the greater part of which are negroes, held as slaves. The commerce is greater than any other country in South America.

Rio Janeiro is the most populous city in South America.

Rio Grande, by Lake Palos, carries on a great trade in hides.

Tejuco is in the interior of the diamond district; and Villa Rica of the gold district.

Guiana.

Noted as being the only portion of South America still under the control of European powers.

It is nearly under the Equator; and on the Demerrara river, and other parts, is said to be unhealthy.

It is claimed by Great Britain, France and Holland.

The land along the coast is level and extremely fertile.

The area is estimated at 150,000 to 160,000 square miles.

Georgetown, the capital of English Guiana, is at the mouth of Demerrara river.

Paramarabo, capital of Dutch Guiana, is eighteen miles from the mouth of the Surinam river.

Cayenne, on an island and strongly fortified, belongs to France.

Buenos Ayres.

Noted for its vast pampas or plains, that feed immense herds of wild cattle, which are taken with the lasso for their hides and tallow.

Buenos Ayres, the capital of Buenos Ayres, or the United Provinces, is one of the largest towns of South America. It is situated about 200 miles from the mouth of the Rio de la Plata; it is well built, and has a large share of commerce. It was founded by the Spaniards as early as 1535.

Paraguay and Uruguay.

Noted for a plant called matte, or Paraguay tea; which is used in several countries of South America in place of the China herb.

Uruguay is noted as the smallest of the South American provinces.

Venezuela.

Scene among the Mountains of Venezuela.

CARACCAS is elevated among the Andes, 3,000 feet above the level of the sea. It was destroyed in 1812, by one of the most awful earthquakes upon record. The number of inhabitants killed is estimated as high as ten thousand. La Guayra, seven miles distant, is its port.

Ca-rac′-cas from the earthquake scarce survives,
Of eighteen-twelve, that cost ten thousand lives;
Three thousand feet she climbs the mount to heaven;
La Guay′-ra is her port, miles distant, seven.
There Mar-a-cay′-bo, to her lake allied,
With Cu-ma-na′, that smiles above the tide.

New Grenada.

Eight thousand feet and seven hundred more,
Stands Bo-go-ta′, o'er New Grenada's shore;
By *Bogota's* rough, rolling tide her state,
Just fifteen miles northeast the *Cataract.*
From Bogota′, southwest, among the mountains,
Proud Po-pay-an′ hears *Cau′-ca's* murmuring fountains,
Upon the northern coast, is Carthagena,
Where roars the *Charib tide* and *Magdalena.*
As Pan-a-ma′ along the south we follow,
Upon the Isthmus, north, is Porto Bello.

Chili.

Val-div′-i-a and Conception, Chili keeps [*Chee′-le*
Along her shores by the Pacific deeps;
And Val-pa-rai′-so opes her harbor wide—
The port for San-ti-a′-go near her side.
Co-quim′-bo of her copper mines may boast;
As Huas′-co, for her silver 's valued most.

SANTIAGO, the capital, is on an extensive plain, fifty miles from sea. Valparaiso has a fine harbor, and is the port for Santiago.

Venezuela.

Noted as the birthplace of Bolivar, and for its llannas or plains, that support large herds of wild cattle; the tallow and hides of which form the chief articles of export. It has an area of over 400,000 square miles.

This province is watered by the Oronoco and its branches. This great river is subject to inundations, which render its shores uninhabitable, but like the Nile of Egypt, deposits a richness and fertilizing quality to the soil.

New Grenada.

Noted as embracing the Isthmus of Darien, and as lying upon two oceans. It has an area of 450,000 square miles. This is the most northern part of South America, and contains some of its sublimest features. The great though natural Bridge of Icononza extends across a crevice or chasm between two perpendicular walls of rock, at the bottom of which flows the torrent of Summa Paz. The arch is 360 feet above the water, and the effect produced by looking down is said to be painful.

BOGOTA, or Santa Fe de Bogota, 8,000 feet above the level of the sea, is the capital.

Falls of Tequendema.

The falls of Tequendema, in the river Bogota, fifteen miles from Santa Fe de Bogota, are among the sublimest in the world. The River Bogota rises 9000 feet above the level of the sea, and makes its way over precipices and through frightful ravines until it comes to the steep of Tequendema, and then plunges 600 feet into a deep and awful chasm.

Chili.

Noted for its delightful climate, and for the Aricaunians, a race of the bravest Indians on the continent; who have never been conquered. Also for being more thickly settled than any other portion of South America.

It is bounded on the north by the Desert of Atacama; on the east by the Andes; south by Patagonia, and west by the Pacific. Its length is over 1,200 miles, and average width about 150.

Peru.

Seven miles from sea, upon a river, narrow,
Lima, fair city, stands—famed for Pizarro;
Her port, Cal-la′-o, beams above the tide;
Famed for its harbor, and well fortified.
Cuzco, where once the Incas held the throne,
Now noted for its Temple of the Sun,
From Lima, stands five, fifty miles, southeast, [550
And in the interior of the province placed.
'Tween this and Lima, on the Lima road,
Gu-an′-ca Vel′-i-ca makes her abode,
Near fourteen thousand feet high o'er the field,
Sublime she sits; her mines quicksilver yield.

Peru.

The ladies of Lima are distinguished for beauty, vivacity, coquetry, love of dress, and admiration. In walking abroad they invest themselves in a cloak called the *manto*, which hides their face from view, so that they are enabled to pass among their most intimate friends without being recognised.

Ladies of Lima.

Interior of the Temple of the Sun, at Cuzco.

Peru was invaded by Francis Pizarro, a Spanish adventurer, in the year 1531. He made the *inca*, or Peruvian king, his prisoner; who offered for his freedom, as much gold as would fill a large room in his palace, piled as high as he could reach; which was no sooner delivered than the blood thirsty Pizarro condemned the innocent king to a cruel death.

At the time of Pizarro's invasion, the Peruvians were fire worshipers.

The empire, according to their tradition, was founded by Manco Capac, in the 10th or 11th century; whose first appearance was on a small island in Lake Titticaca. He avowed himself and wife to be children of the sun, sent down to enlighten and civilize the nations. He taught the men agriculture and other useful arts, whilst his wife instructed the women in spinning, weaving and other domestic affairs.

Peru is noted for its rich mines of gold and silver; and for having been the seat of one of the most civilized nations in South America. It is bounded on the north by Equador; east by Brazil; south by Bolivia; west and southwest by the Atlantic. The area is reckoned at 500,000 square miles. Rain seldom or never falls on the coast of Peru, south of Cape Blanco. The earth, from June to November, is wet with heavy dews and fogs. Cinchona or Peruvian bark, so valuable in medicine, is obtained from this country.

The coast of Peru is unfavorable to navigation, and with the exception of Callao, has no good harbors. The surf and breakers are so tremendous that no ordinary boats can land, or reach the shore. To remedy this the natives inflate with air large bags, made of skins; of which a sort of raft is constructed. called the *balsa*; with this they are enabled to load and unload vessels.

Equador.

Quito and the Andes.

Quito is celebrated in Geography for its sublime elevation, and unparalleled and unequalled serenity of climate. It is situated among the Andes, 9000 feet above the level of the sea; and though under the equator, has an atmosphere fraught with eternal April.

Qui-to is seated on her mountain throne,
Nine thousand feet, and in a burning zone;
Perpetual spring around those summits keep,
And pure the zephyr in its gentle sweep.
And Guayaquil, chief port of Equador,
From this is south, along the surf beat shore.

Bolivia.

La Plata, in Bo-liv'-i-a, we see,
Much noted for its splendid scenery.
As Po-to-si' is known for silver mines, [*po-to-see*
La Paz', from Potosi, northwestward shines.

La Plata, or Chuquisaca, the capital of Bolivia, is noted for the splendid scenery found in its vicinity.

Patagonia is noted as being the coldest and most southern portion of South America; also for the gigantic stature of the Indians that inhabit it. It is bounded on the north by Buenos Ayres; east by the Atlantic; south by the Straits of Magellan, and west by the Pacific. Its area is estimated at 300,000 square miles.

The natives of this country are probably the tallest and most gigantic in the world; their average hight being six feet and a half.

PRONUNCIATION.

Buenos Ayres,	*Bwa'-nos Ayres.*	Guyaquil,	*Gau-a-keel.*
Quito, - -	*Ke-to.*	Rio Janeiro,	*Rio Jan-e'-ro.*
Lima, - -	*Le'-ma.*	Cuiaba,	*Ki-a'-ba.*

Lima, the capital of Peru, is situated in a rich vale, watered by the Rimac. It is celebrated mostly for its founder, Francis Pizarro. The streets are wide and regular. In the middle of the town is the Great Square, one of the largest and finest in America; in the middle of which is a large fountain having a bronze statue of Fame in the center.

Cuzco, 550 miles southeast of Lima, is noted as having been the capital of the *Incas*, or Peruvian kings; and also for containing the remains of a magnificent temple of the Sun, the interior view of which is given on the foregoing page.

Guanca Velica, between Cuzco and Lima, has the highest elevation of any town on the globe, excepting Potosi, which is probably as high. Its mines of quicksilver were once sources of great wealth.

Equador.

Noted as lying under the equator, and for containing some of the loftiest peaks of the Andes. It is divided into three parts; Equador, Guayaquil and Assuay.

Equador consists of table lands. The climate is the finest in the world and resembles spring, the year round. It is called the Evergreen Quito.

Bolivia.

Named in honor of General Bolivar, the liberator of most of South America. Noted for containing the loftiest peaks of the Andes; also for the celebrated mines of Potosi.

Its area is estimated at 400,000 square miles. It is rough and mountainous.

Sorato, the highest peak of the Andes, towers aloft to the highth of 25,380 feet, or nearly five miles high. Illimani, the second highest peak of the Western Continent, is 24,350 feet.

The Condor is the largest bird of the air; and measures, when its wings are extended, sixteen feet. It soars aloft above the highest peaks of the Andes, and descends to the vales only in pursuit of prey.

The Condor.

Potosi, elevated 13,265 feet above the level of the sea, is celebrated for its rich silver mines. The city is situated on the west side of Mount Cerro de Potosi, that contains the precious metal. They were first discovered by an Indian, who, pursuing a lama up the declivity, caught hold of a bush, which being torn up by the roots, revealed a solid mass of silver.

The produce of these mines from 1556 to 1800, amounted to 823,950,508 Spanish dollars.

La Paz, northwest of Potosi, is the principal town of Bolivia. The lofty summits of Sorato and Illimani, are seen from this place crowned with eternal snow.

EUROPE.

Europe is noted as the smallest, but wealthiest, most thickly populated, and most powerful and enlightened of the general divisions of the globe.

Coliseum at Rome.

"While stands the Coliseum Rome must stand,
When falls the Coliseum Rome must fall."

Capes.

Far to the north, where roars the Arctic Sea,
In seventy-one, (71) *North Cape* is known to lay.
The *Naze* of Norway, and *La Hogue* of France,
One south, one north, above the waves advance.
To the north of Spain *Cape Or'-te-gal* may fare,
While north and westward, stands *Cape Fin-is-terre'*.
Southwest of all, *St. Vincent* shines afar,
As near Gibraltar, stands *Cape Traf-al-gar*,
And *Ga'-ta* south, *Pa'-los* southeast of Spain,
While on the east, *St. Martin* finds a reign.
South of Sar-din'-i-a, *Spar-ti-ven'-to* peers,
While north of Corsica, *Cape Cor-so* rears.
And Italy bids her *Spar-ti-ven-to* smile,
As *Pas-so* 's south of Sicily's fair isle;
And south of Greece *Cape Mat-a-pan'* behold,
Where last of all, *St. An-ge-lo* 's enrolled.

Rivers.

RIVERS OF IRELAND.

From E-rin northward runs the *River Foyle*,
With *River Bann*, whose flashing waters boil;
In George's Channel, eastward rolls the *Boyne*, [boin
Where *Lif-fey* pours and *Slaney's* billows shine;
Southward, in seven west, behold the *Bar-row*;
Blackwater then, and *Lee*, in channels narrow.
From Allen, Rec, and Derg—three sylvan lakes,
Southwesterly her waves the *Shan-non* takes.

RIVERS OF SCOTLAND.

From Scotland, to the North Sea, runs the *Tweed*,
'Tween Scot and Englishman she finds a bed;
Then comes the *Forth*, the *Tay*, and flashing *Dee*,
Still further north, with *Don* and rolling *Spay*.
As southward, on her western coast we roam,
The *Clyde* first greets us, then the *Ayr* and *Doon*;
The *Dee* and *Nith* with *An-nan* southward pass,
With *Esk* and *Lid-dell*, bound to Solway Frith.

RIVERS OF ENGLAND.

The *Tyne* and *Tees* come first then with the *Humber*,
Ouse, *Air* and *Trent*, branches just three in number;
These with the *Thames*, from Britain's eastern coast,
Are hurried on, and in the North Sea lost.
To the British Channel westward rolls the *Severn*,
As *Mersey* to St. George's Strait is driven.

OF PORTUGAL AND SPAIN.

Northward of all, upon the Spanish shore,
The *Min'-ho's* waters to the Atlantic pour;
And here the *Due'-ro* and the *Tagus* drain,
With *Guar-di-an-a*, Portugal and Spain;
Here *Gua-dal-quiv'-er* An-da-lu-sia sends,
And Ebro in the Mediterranean ends.

OF FRANCE.

Ga-ronne' and *Loire*, in Biscay's Bay are thrown,
And the Gulf of Lyons drinks the flashing *Rhone*,
The English Channel swallows up the *Seine*,
That runs from France where Paris holds her reign.

RHINE AND BRANCHES.

The *Rhine* from Switzerland makes her first advance,
Then northward turns—kissing the shores of France;
In Germany and Holland then she's seen,
Taking from Germany *Mo-selle'* and *Mayne*; [main]
From Holland, *Meuse*, that moistens Belgium's earth,
Coming from France—the province of its birth.
The *Ems*, the *Weser* and the *Elbe* are hurried
O'er German shores, and in the North Sea buried.

OF RUSSIA AND PRUSSIA.

From Prussia to the Baltic, *O-der* glides,
As *War-tha* river in her channel hides.
Vis-tu-la there, with *Neimen's* waters tread,
By Prussia nurtured and by Russia fed.

OF LAPLAND AND SWEDEN.

From Lapland, *Au'-nis* seeks the Bothnia's roar;
Where *Tor'-ne-a*, *Ca-lix* and *Lul-le'-a* shower,
With *River Dal*, from Sweden's wintry shore.

OF RUSSIA.

Du-na from Russia on to Riga storms,
La-do'-ga Lake to Finland, *Ne-va* turns.
Onei'-ga, *Dwi-na*, and the *Mez-en* sweep,
O'er Russia's shores into the White Sea deep.
From the same fields, with all her ice in motion,
Pet-cho'-ra runs into the Arctic Ocean.
Vol-ga and *U-ral* seek the Caspian Sea;
The last is Europe's eastern boundary.
And *Kuban's* waves, the Black Sea's waters greet,
As *Don* and *Donec*, near the Azof meet.
And *Dneiper*, *Bog* and *Dneister*, all are lain [*neister*
From Russia, to the Black Sea's raging main.

OF NORWAY.

The *Glom-ma* rolls her down the Norway coast,
And in the Strait of Cat-te-gat is lost.

DANUBE AND BRANCHES.

Here *Dan-ube* comes, the tide that Swiss and German,
And Austrian and Turk, all hold as common.
A northern branch the *River Pruth* is seen,
The boundary line 'tween Russ and Ottoman.

RIVERS EMPTYING INTO THE ARCHIPELAGO, ADRIATIC AND MEDITERRANEAN.

Vardur and *Struma* with *Marissa* go
From Turkey to the Archipelago;
Narema westward with the rolling *Drin*,
Foams where the Adriatic waters grin.
From Italy here comes the river *Po*;
While westward, *Arno* and the *Tiber* flow
In Mediterranean, with the *Vol-tur-no'*.

TOWNS AND COUNTRIES.

British Empire.

The most powerful, and with the exception of China, the most populous empire on the globe.

It embraces England, Ireland, and Scotland, with the principality of Wales, Gibraltar and the Island of Malta, the greater part of Hindoostan and the Island of Ceylon; Sierra Leone, and several forts in Guinea, the Cape of Good Hope, St. Helena and Mauritius, British America and Honduras, Jamaica, Barbadoes, Trinidad and several of the West India Islands, English Guiana, Australia, Van Dieman's Land, New Zealand and other islands in different parts of the world.

The population of the whole is estimated at 160,000,000, or eight times that of the United States.

Her great power and bulwark is her vast navy, that once outnumbered the combined navies of the world.

Her commerce is greater than any other nation. The merchant vessels are upwards of 27,000 in number; of the burden of 3,050,000 tons. These are navigated by 180,000 seamen.

England.

Windsor Castle

Windsor Castle, on the Thames, is one of the royal residences of the Queen. It was built by William the Conqueror.

On England's shores, London is first surveyed,
The queen of towns in commerce, arts and trade.
And Liverpool upon the *Mersey* lay,
The port for Ireland and America.
Manchester, east of Liverpool we enter,
Of cotton manufacturing 'tis the center.

England.

The southern part of Great Britain; the seat of the British Empire; and noted as the most important state in Europe, and as exerting the greatest influence upon the destinies of the civilized and uncivilized world.

England abounds in beautiful and interesting scenery. Its agriculture is superior to that of any country of Europe. It is the most extensive manufacturing country in the world. In no part of the world is wealth more unequally distributed. The government is a limited hereditary monarchy.

The national debt is $4,000,000,000; the interest of which is $160,000,000 annually, or six times as much as the whole expenditures or the United States' government.

Wales.

A rough, rugged, and mountainous country, west of England; noted for its mines of lead, iron, copper, and coal, and as giving the title of "Prince of Wales" to the English sovereign's eldest son.

LONDON, the capital, on the Thames, 60 miles from its mouth, is the largest city on the face of the globe both in extent and population. It covers about 25 square miles in area. Among its public

For woolen factories, Leeds first is classed,
As Birmingham for hardware 's unsurpassed.
Sheffield, for cutlery may wear the crown,
For stockings, Nottingham and Leicester's known.
For watches Coventry is famed the while;
Swan-se-a, for copper works comes in the file.
Bristol is noted for her wells so hot;
As Hull, upon the *Humber* finds a spot.
Portsmouth is found upon the southern coast,
With Plymouth, for her sea wall noted most.
Windsor and Warwick, for their castles named;
Oxford and Cambridge, for their schools far famed;
Scar'-bo-rough, Bright'-on, Cheltenham, and Bath;
Chief watering places, all beneath my path.

Scotland.

Throned on her hills, for science first in place,
Is Edinburgh, Scotia's metropolis.
Leith is her port, beside the *Forth* we learn,
Upon whose shores is the famed Bannockburn.
Proud Stirling here displays her warlike dress,
As north on *Murray's* banks is In-ver-ness'.
Fair Ab-er-deen, between the *Dee* and *Don*,
For building ships has gathered much renown.
Glasgow is found upon the *River Clyde;*
Greenock, her port, is twenty down the tide.
Paisley from Glasgow, west miles eight or nine,
Is noted for its cotton goods so fine.
On Scotia's eastern shore, behold Dundee',
Spinning her canvas on the banks of *Tay*.
Known as the abode of Scotia's ancient kings,
Perth on the *Tay*, linen and cotton spins.

Ireland.

Seven miles is Dublin from the Irish Sea,
On *Liffey's* banks, she rules o'er bond and free.
Gal'-way is west, as Cork is to the south,
While Limerick lives far up the *Shannon's* mouth.
Belfast north-east may at her linen toil:
As Londonderry lives upon the *Foyle*.

Spain.

Madrid, the capital, on table-lands,
In the interior of the kingdom stands.
Fair Barcelona smiles above the sea;
In manufactures and in commerce free.

buildings, are St. Paul's Church, the Tower, Westminster Abbey, and Bank of England.

The bridges of London, are works of great labor and expense.

The Tunnel, under the Thames, is one of the great achievements of art.

Sheffield has the most noted cutlery in the world. Bristol is noted for hot springs.

Plymouth is noted for its immense breakwater, that cost $5,000,000.

Scotland.

North of England, separated by the Grampian Hills into two parts; North and South, or the Highlands and the Lowlands.

It abounds in wild and sublime scenery.

The Highlands are mountainous, the Lowlands more level, and better adapted to tillage.

The Highlanders are brave, hospitable and independent, and possess a rude and lawless kind of character.

EDINBURGH is the capital.

Leith is the port of Edinburgh.

Bannockburn, on the Forth, is noted for the victory of Bruce over the army of Edward II, of England.

Stirling, on the same river, is a strongly fortified town.

Dundee, in the eastern part of Scotland, on the River Tay, is noted for the manufacture of canvas.

Ireland.

Ireland, called the "Emerald Isle," "Green Erin." A large island west of England. The native land of the Irish.

The surface is uneven, but not mountainous. Bogs and marshes cover one tenth of its surface. The peat bogs supply the fuel.

Barley, oats, wheat, flax, &c., are extensively raised.

Potatoes constitute the chief product, and before the potato rot of late, formed the principal food for the poor. Ireland has been sorely oppressed by its English rulers for centuries past.

The Irish are quick witted, sanguine, warm-hearted and hospitable, but prodigal and passionate.

Four-fifths are Catholics, and the remainder Protestants.

For a few years past, Ireland has been in a starving and deplorable condition, from the failure of her crops.

DUBLIN, the capital, is on the Liffey, seven miles from its mouth.

Galway is in the western part of the island.

Limerick is on the Shannon, in the interior.

Belfast is noted for the manufacture of fine linens.

Spain.

Noted for its salubrious climate and picturesque scenery, and as having been one of the leading powers of Europe; but now one of the most feeble and unimportant.

The soil is fertile, but poorly cultivated. Her

There Al-i-cant' and Car-tha-ge-na rest,
Of Mediterranean ports the last is best.
There Mal'-a-ga for fruits and wines is known,
On Andalusia's southern shores her home;
Cadiz, her bulwarks o'er the Atlantic rears,
North-west the Straits, where strong Gibraltar peers.
North-west of all Co-run'-na lives, the station,
For ships of Britain and the Yankee nation.
Fer-rol' stands here, where Spain her navy gathers,
Near where Cape Ortugal the dark sea weathers.
On the *Bay of Biscay*, whence the wool of Spain
Exported is, Bil-bo'-a finds a reign.
Valencia 's noted for her silks so fine;
Xeres is known quite well for sherry wine. (ze-res)
Se-ville', Grenada, and Cor-do'-va lower,
All splendid cities once, of Moorish power.

Portugal.

Lisbon, with wines and fruits where Tagus fills
The Atlantic bowl, is throned on several hills.
St. Ubes, south-east of Lisbon makes a halt,
And from the sea-wave manufactures salt.

Coimbra.

Coimbra, 120 miles north-east of Lisbon, is noted for its University. The palace of the University, once the residence of the kings, is one of the finest buildings in the place.

And north of Lisbon, next Co-im'-bra see,
Much noted for her university.
Oporto, on the *Duero* makes resort,
Known the world over for a wine called *Port*.

France.

Paris, in gardens, palaces and pride,
Fashions and gaiety, is not outvied.
Lyons in manufacturing takes her throne,
Just at the junction of the *Rhone* and *Saone*.
Mar-seilles, in commerce is by none surpassed,
Bordeaux in wines, much money has amassed;

commerce and manufactures are in a neglected state. It is separated from France by the Pyrennees, among which are found numerous monks and hermits.

The richest portions of America once belonged to Spain, though Cuba and Porto Rico are all that now remain.

The other colonies are the Philippine, Caroline, and Ladrone Islands in the Pacific, and the Canary Islands in the Atlantic Ocean. These are her chief sources of national revenue.

Madrid, the capital, is situated on the table-lands in the interior.

Barcelona, on the Mediterranean, is noted for commerce and manufactures.

Carthagena is noted as being the best port on the Mediterranean.

Malaga, in the province of Andalusia, is noted for its rich wines and delicious fruits.

Cadiz is strongly fortified.

Corunna is noted for the battle of Corunna, between the French and English, and as the port or station for packets of Great Britain and the United States.

Ferrol is noted for a naval station; Bilboa for its commerce in wool; Valencia for its silks; Xeres for sherry wine; Seville, Grenada, and Cordova, as important Moorish cities.

Gibraltar.

The Promontory of Gibraltar constitutes the strongest fortress in the world. It is three miles long, half a mile wide, and 1400 feet high. It commands the entrance to the Mediterranean. It is in the possession of Great Britain.

Portugal.

Portugal was once the most commercial state of Europe, but is now reduced to insignificance.

The climate is remarkably mild and healthy.

Agriculture, manufactures, education, and improvements of every kind, are in a backward condition. The only productions of importance are wine and salt.

Lisbon, the capital, is situated on the Tagus.

St. Ubes is noted for salt; Coimbra for its university; Oporto for the production of Port wine.

France.

Noted for the important part she has acted in the affairs of Europe, and as having lately become a Republic, the only one of consequence on the Eastern Continent.

The climate of France is mild and salubrious.

Havre, fair port of Paris on the *Seine*, (hav'-r)
Tou-lon' and Brest, as naval stations reign.
Roche-fort' and l'Orient on Biscay's Bay, (lo-re-ong)
Are naval stations too, where ships may lay.
Bay-onne', near Spain, for bayonets long known,
Cal-ais', that oft has bowed to England's throne.
Bou-logne, south-west from this her station finds,
Rou-en', upon the *Seine* her cotton spins.

Russia.

St. Petersburgh is 60 from the Equator, (60°)
By *Neva's* banks she rules, and none is greater.
Cron-stadt, a naval post where Finland boils,
West from St. Petersburgh just twenty miles.
And south of Petersburgh, is Nov-go-rod':
Though fallen now, she boasts of royal blood:
The proudest city once of all the north,
Godlike in power, imperial in worth.
Pol-to'-wa, known for Charles the XII of Sweden;
Whose fate we learn when history we're reading.
O-des'-sa, by the *Black Sea*, takes her seat,
And from this place exports the Russian wheat;
And Ni-ca-la-jef' there joins the catalogue,
A naval station, seated on the *Bog*.
Known for her palaces, and for her bell,
Moscow in the interior may dwell.
As north of all Archangel's lair is made,
Riga on *Riga Gulf* may boast her trade.
And by the *Caspian* of the Russian clan,
On *Volga's* southern bank, is Astracan.
Known for her battlements, and for her wall,
Of Poland, Warsaw reigns the capital.

The Russians, in general, are robust, well shaped, and of pretty good complexions. The dress of the higher ranks are after the French and English fashions; and all wear a covering of fur for six months of the year. Persons of both sexes wear a cross on their breasts, which is put on when they are baptized, and never laid aside while they live.

The following are the Sovereigns of Russia, showing the years of their accession to power:

Peter the Great, ascension in	1696
Catherine I.,	1725
Peter II.,	1727
Anne,	1730
John,	1740
Elizabeth,	1741
Peter III.,	1762
Catherine II.,	1762
Paul,	1796
Alexander,	1801
Nicholas,	1825

The vineyards yield 850 million gallons of wine annually, and occupy five million acres of ground. The principal colonies are Algeria, Senegal, and the Isle of Bourbon in Africa; Martinique and Guadaloupe, in the West Indies; French Guiana in South America; and Pondicherry in Asia.

PARIS, the gay capital of the French Republic — the paragon of fashions for the world — is on the Seine.

Lyons, noted for manufactures, is at the junction of the Rhone and Saone.

Marseilles is noted for commerce; Bordeaux for wines; Havre as the port of Paris; Toulon, Rochefort, and Brest as naval stations; Bayonne as the place where bayonets were first used; Calais as having repeatedly been in the possession of Great Britain; Rouen for cotton manufacturing.

Russian Empire.

It comprises nearly one half of Europe, one third of Asia, and a part of North America. It extends half way round the earth, and comprehends one seventh of the land's surface. It is, generally, a level country, and its characteristic features are vast plains and majestic rivers.

Russia.

Russian Europe is noted for its great power and importance.

The inhabitants are Russians, Poles, Finns, Tartars, and Cossacks; the latter form a most efficient part of the army of Russia.

The Emperor is at the head of the church, and is styled the Autocrat of all the Russias.

The military force, or army of Russia, is the largest by far of any in Europe, and is a great object of terror and anxiety throughout all the Eastern Continent: it amounts to nearly 1,000,000 men. The naval force consists of 300 vessels, 50 of which are ships of the line.

The great body of the Russians is divided into two classes: nobles and slaves. The former live in great splendor; the latter are the property of the nobles or the emperor.

Twenty-two millions of serfs or slaves are said to be owned by the Autocrat himself.

ST PETERSBURGH, the capital, on the banks of the Neva, and 60 degrees from the equator, is one of the most splendid cities in the world.

Novgorod, though now in a decayed state, was once the seat of a great republic.

Poltowa is remarkable. in history, for a great battle fought between Charles XII of Sweden, and Peter the Great of Russia, in which the latter gained a complete victory.

Moscow was burnt by the Russians, in 1812, to prevent its falling into the hands of the French. It was celebrated for its mammoth bell, the largest ever cast, the weight of which was upwards of 150 tons.

Lapland

Noted as being the most northern country of Europe. It is owned by Russia and Sweden.

The inhabitants are called Laplanders or Lapps. They are a simple, inoffensive race; strictly honest; and live to a great age. In stature they never exceed five feet.

Republic of Cracow.

Cra′-cow, beside *Vistula*, takes her post,
Known for the mound of Kos-ci-us′-ko most.

Sweden.

Stock-holm, in Sweden, is the brightest star,
On seven small isles, 'tween *Baltic* and *Ma-lar′*.
At *Gotha's* mouth, whose fountain head is Wenner,
Fair Gottenberg spreads her commercial banner.
Of Fah-lan's copper mines, go read the story,
And then, for iron, look at Dan-e-mo-ra.
In fur, Tor-ne-a trades; she's north of all;
Carls-cro-na, south, a naval station call.

Norway.

Bridge and Mountain torrent in Norway.

Christiana on the Norway coast is laid;
Iron and lumber is her wealth and trade.
Upon the western coast, is seated Bergen;
In lumber, tar and fish, her commerce urging.
Dron-theim′ is north of this, along the flood—
Of Norway's ancient kings, 't was once the abode.

Prussia.

Berlin, of Prussia, stands upon the *Spree*,
A branch of *Elbe*, of royal pedigree.
Bres-lau is found far up the river *O-der*,
And known for linens, near the Polish border.
And Konigsburg is seated on the *Pre-gal*,
Whose place or rank, in days gone by, was regal.
Next, Dant-zic, on the *Vis-tu-la*, we greet,
Great mart of Poland, for exporting wheat.

Republic of Cracow.

Noted for a mound raised to the memory of Kosciusko, which is 300 feet in highth, and 275 feet in diameter at the base.

Sweden.

Noted for its numerous lakes. It is a level country, with the climate of Canada East, and has about 2,400 miles of sea coast.

It has valuable mines of iron and copper.

Hardly one thirtieth of the land is tillable.

The higher classes of the Swedes are intelligent, brave and hospitable; but luxurious and ostentatious.

The peasants are simple, kind and strictly honest.

The complexion of the Swedes is ruddy; the hair flaxen; and their beards and moustaches have been described by travelers as almost white, and in beautiful keeping with their blue eyes and rich complexions.

STOCKHOLM, between Baltic and Malar, is the capital.
Danemora has the best iron in the world
Fahlan is noted for copper, and Tornea for fur.

Charles XII. of Sweden, was one of the greatest of modern warriors. He came to the throne in 1697, at the age of fifteen. In his seventeenth year, he fought the combined armies of Russia, Poland and Denmark, and gained over them a decisive victory. In his first battle when he heard the hissing of the bullets about his ear he exclaimed, in a rapture, "That shall be my music."

Norway.

Noted for its rugged mountains, cold climate, gigantic pines, and for the terrific whirlpool on its coast, called the Malstrom.

It is united with Sweden under one government, though each state enjoys its own constitution, its own laws and legislature.

From 1380 to 1814, it was united to Denmark; but since that time it has formed a part of Sweden.

It is one of the most mountainous countries in Europe, and abounds with romantic and sublime scenery.

CHRISTIANA, the capital, is noted for iron and lumber.

Bergen, upon the western shore, carries on a great trade in lumber, tar and fish.

Prussia.

Noted for its rapid rise from a small state to one of the first powers of Europe.

It was formerly an electorate of Germany; Brandenburg the basis: East and West Prussia were first added; Silesia was wrested from Austria; Posen from Poland; and a part of Pomerania from Sweden; and Saxony, Westphalia, Clevesburg and the Lower Rhine, were added in 1815.

The principal rivers are all navigable.

Amber is found on the shores of the Baltic.

The army is the best disciplined in Europe.

Co-logne, upon the *Rhine*, with Dutch may trade,
A water, called *Cologne*, she long has made.
As Frankfort on the *Oder* keeps her fairs,
On *Elbe*, is Mag-de-burg, prepared for wars.
As Luther lived in Wittenberg—in Thorn,
By *Vis-tu-la*, Copernicus was born.
Aix la Cha-pelle and Til-sit both, we find,
Are for important treaties borne in mind.

Austria.

On *Danube's* banks, o'er Austria stands Vienna,
Upon a fertile plain, she rules o'er many.
Prague rules Bohemia, on the tide *Moldau;*
Her bulwarks frown upon the fields below.
North of Vienna Aus'-ter-litz appears,
And of Napoleon's victory wears the scars.
By the *Adriatic Gulf* is throned Tri-este,
Well fortified, of Austria's ports the best.
And near the *Adriatic*, 'mong the number,
I'-dri-a, for quicksilver mines, remember.

Buda and Pesth.

Buda, on the west bank of the Danube, is connected with Pesth, on the eastern bank, by a bridge of boats.

At Bu-da's baths and palaces, now look—
By *Danube's* western bank, upon a rock;
With this, connected by a bridge of boats,
Pesth, on the eastern bank, her trade promotes.
Schem'-nitz and Krem'-nitz, 'mong the mountains [old,
Well known for mines of silver and of gold.
To-kay', for wines; as Presburg, well you know,
Was Hungary's capital, some years ago.
Lem-berg, for inland trade; of Polish birth,
Wie-licz'-ka from her salt mines draws her worth.

German States—Bavaria.

Bavaria waves her banners by the *I'-ser,* [*e-ser*]
Bavaria's capital, she's known to be, sir.

Its system of common school education is considered the best in the world.

BERLIN, the capital, on the Spree, a branch of the Elbe, is one of the most splendid cities in Europe.

Konigsburg was once the capital of the whole kingdom.

Frankfort, on the Oder, is noted for fairs.

Magdeburg, on the Elbe, is strongly fortified.

Wittenberg was the residence of Luther.

Thorn was the residence of Copernicus.

Aix la Chapelle and Tilsit are noted for treaties.

Austria.

One of the most important states of Europe; one third larger than France, and twice as large as Great Britain and Ireland.

It is richer in minerals than any other European state.

The Archduchy of Austria is the original basis of this Empire. Hungary was obtained in 1438, by marriage; Gallicia in 1792, by the dismemberment of Poland. The Italian provinces were annexed in 1815.

VIENNA, on a fertile plain, and situated upon the Danube, is the capital.

Prague, on the Moldau, a strongly fortified town, is the capital of Bohemia.

Austerlitz, north of Vienna, is noted for a great victory of Napoleon over the Austrians.

Trieste, by the Adriatic, a strongly fortified town, has the best port in Austria.

Buda, noted for baths, on the Danube, is connected with Pesth, on the eastern side, by a bridge of boats.

Wieliczka, a town of Poland, is noted for salt mines.

Kremnitz.

Kremnitz and Schemnitz, among the mountains of Hungary, are noted for gold and silver mines.

German States.—Bavaria.

Bavaria, in the southeastern part, is, next to Austria and Prussia, the most important state in Germany.

The Black Forest and the Alp, two masses of mountains, form one principal feature of this state. They are bleak regions, with little wood or verdure.

Agriculture is in a backward state, and manufactures have been neglected.

BAVARIA, on the Iser, is the capital.

For watches known, inventions, toys and books,
At Nuremberg, the traveler often looks.
Blenheim, on *Danube*, and the Ho-hen-lin-den,
From Mu'-nich, east, the war-horse once reclined on.

Saxony.

For her picture galleries known, now look at
Beside the *Elbe*, as Saxony, she rests in. [Dresden,
In fairs and commerce, let fair Leip'-sic reign,
As Meis'-sen, on the *Elbe*, makes porcelain;
And Frey'-berg, by two hundred mines surrounded,
Has there a mining institution founded.

Hanover.

Upon the *Leine*, from the western shores, just over,
Well fortified and strong, is fair Hanover.
Up the same tide, has Gottingen her post,
Known for her university the most.
A North Sea port, for vessels great and small,
Em'-den is on the *Ems*, northwest of all.

Wirtemberg.

Stutt'-gard of Wir'-tem-berg, as first may shine,
Upon a branch of the fair flowing *Rhine*. [dress,
On *Danube's* banks, Ulm wears her shining war-
O'er Europe, noted for a mighty fortress.

Baden.

Carls-ru'-he, near the *Rhine*, rules over Ba'-den,
Whose shores, with Man'-heim, farther north, are laden.

Small German States.

Mentz, on the river *Rhine*, invented printing;
Of war, her bulwarks seem to be a hinting.
Weimar, the capital of Saxe Weimar,
Has been the abode of many a learned dreamer;
Je'-na, southeast of this, her place may fix,
Remembered for the fray of eighteen 'six. [1806

Free Cities.

Frankfort, known for her Federative Diet,
Lives on the *Mayne*—a place of fairs and quiet.
Hamburg, upon the *Elbe*, has fixed her station,
Where vessels come to trade, of every nation.

The battle of Hohenlinden, where Moreau gained one of his great victories, is commemorated by Campbell in a sublime and glowing song, of which the following is an extract:

On Linden when the sun was low,
All bloodless lay the untrodden snow,
And dark as winter was the flow,
Of Iser, rolling rapidly.
But Linden saw another sight,
When the drum beat at dead of night,
Commanding fires of death to light,
The darkness of her scenery.

Saxony.

The smallest kingdom in Europe, though the people are the best educated.

Agriculture and mining mostly form the occupation of the inhabitants.

The Saxon sheep are noted for their fine wool.

Commerce and manufactures are extensive.

DRESDEN, on the Elbe, is noted for its picture galleries.
Leipsic is noted for fairs.
Freyberg, in the center of the mining district, is noted for its mining institution.

Hanover.

Hanover became a kingdom in 1815.

It is mostly an extensive plain, with gentle rising grounds, and nearly destitute of mountains. The Hartz Mountains are rich in mines, which are extensively wrought.

STUTTGARD, on a branch of the Rhine, is the capital.
Emden on the Ems, is the principal port for the North Sea trade.
Gottingen, on the Leine, is noted for its university.

Wirtemberg.

Noted as being the best cultivated part of Germany.

Erected into a kingdom in 1806.

Its mountains are rich in minerals.

Baden.

A narrow but fertile plain on the east side of the Rhine.

CARLSRUHE, is the capital.

Small German States.

HESSE DARMSTADT consists of three separate districts; two north, and the other south of the River Mayne.

SAXE WEIMAR is noted for its high rank in literature and the arts.

Jena is noted for a great battle between the French and Prussians, where Napoleon gained one of his greatest victories.

Free Cities.

Mentz, strongly fortified, is noted as being the place where printing was invented.

Seated upon her western banks, is Bremen,
Noted for commerce, and, of towns, a freeman.
Lu-bec', though in obscurity immersed,
In the famed Hanseatic League, was first.

Switzerland.

The Devil's Bridge.

The Devil's Bridge, in Switzerland, is built over the Reuss, a foaming, rapid torrent, that empties in lake Lucerne, after passing through the canton of Uri. The sensation produced by looking from the top, is giddy and sublime; and the roar of waters almost deafening.

Of Switzerland's towns, Berne stands upon the [Aar;
Lau-sanne, upon *Geneva*, has her fare;
These two, with Zurich, on her lake or sea, [*zu'-rick*
Are noted for their splendid scenery.
Known as the place where paper first was made,
And for her school, Basle on the *Rhine* is laid. [*bale*
Scauff-hau'-sen, for a cataract of the Rhine;
Lu-cerne', where towers the forest tree, sublime.
Known for her council, of religious make,
Constance, northeast of all, is by the *Lake*.

Belgium.

Brussels.

Brussels, in Belgium, is noted for carpets, lace, camblets, &c.

Brussels, in Belgium, on a branch of *Scheldt*, [*skelt*
In carpets, lace and camblets, long has dealt.

The four free cities of Germany are all that remain of the Hanse or imperial towns; which once numbered eighty-five of the most commercial cities in Europe.

Frankfort, the capital of Germany, is noted for fairs, and as the place where the German or Federative Diet, or Congress assemble.

Switzerland.

One of the most mountainous countries of Europe. Noted for its sublime and romantic scenery.

It is divided into twenty-two cantons, which are each independent republics, united together for the purpose of common defense.

The Swiss mountaineers are noted for hospitality and love of liberty.

The country abounds with lofty mountains, covered with perpetual snow; glaciers, or lakes of ice; torrents that roar and foam down the rocks; and avalanches, or immense masses of snow.

Berne, Lucerne and Zurich are the capitals of Switzerland; they are noted for the grandeur of their scenery.

Basle, noted for its university, and as the place where paper was first made, is on the Rhine.

Schauffhausen is near the celebrated cataract of the Rhine.

Constance, by Lake Constance, is noted for its ecclesiastical council.

The Cavern of the three Tells.—The three founders of the Helvetic confederacy are thought to sleep in a cavern near the Lake of Lucerne. It is supposed that if Switzerland is ever enchained, they will arise and vindicate her rights.

When Uri's beechen woods wave red,
In the burning hamlet's light;
Then from the cavern of the dead,
Shall the sleepers walk in might.
With a leap like Tell's proud leap,
When away the helm he flung,
And boldly up the steep
From the flashing billow sprung

They shall wake beside the forest sea,
In the ancient garb they wore,
When they linked the hands that made us free,
On the Grutli's moonlight shore.
And their voices shall be heard,
And be answered with a shout,
Till the echoing Alps are stirred,
And the signal fires blaze out.

Mrs. Hemans.

Belgium.

Noted for its fertility; its high state of cultivation; and for its being the most thickly populated of any country of Europe.

The Belgians were formerly called Flemings.

Belgium once belonged to Austria, and then to France. In 1815 it was united with Holland It became a separate State in 1830, when Leopold took the throne.

Brussels, the capital, is noted for its carpets, lace, camblets, &c.

From Brussels, north, in miles, just twenty-five,
Antwerp, upon the *Scheldt*, her trade may drive;
Of her cathedral, there is much renown,
That climbs the heavens in feet, four, forty-one.(441)
Ghent, for a treaty known, we next will scan,
Just thirty miles southwest of Amsterdam.
Nine miles from Brussels, south, is Waterloo,
Where met Napoleon his overthrow. [quarters.
Liege, known for firearms, makes the *Meuse* her
Mech'-lin for lace, and Spa for mineral waters.
Tour'-ney and Mons along the French frontier,
Safe in their battlements, need nothing fear.

Denmark.

And Copenhagen stands on Zealand isle;
As, by the *Elbe*, Al-to'-na reigns the while.
From Copenhagen, north, on Zealand's shore,
Where vessels pay their toll, is El-si-nore'.

Holland.

In proportion to its extent, Holland is one of the most populous districts on the globe.

The Dutch were, at a former period, the most flourishing and greatest commercial people on the globe.

The foreign territories belonging to Holland are chiefly in the East Indies, and include part of the islands of Java, Sumatra, Banda, Borneo, Celebes, Gilolo, and Timor, also the Moluccas or Spice Islands. In South America, Surinam or Dutch Guiana. In the West Indies, the islands of St. Eustatia, Curacoa, Saba, and part of St. Martin's. In Africa, several forts on the coast of Guinea.

In Holland, near the coast, Hague makes her claim;
As, thirty miles northeast, is Amsterdam;
The last stands on an arm of Zuy-der Zee,
Known for canals, where boats pass merrily.
As Rotterdam is seated on the *Meuse*; [muce
Harlem, by *Harlem Lake*, of flowers makes use.
As a naval depot, next, remember Flushing,
U-trecht', for peace, where river *Rhine* is blushing.

The States of Italy.—Sardinia.

Much noted for her silks, beside the *Po*,
Tu-rin rules o'er Sardinia you know. (tu-reen')
Gen'-o-a, built upon a mountain's side,
Still of Columbus makes her boast and pride.
Here, Al-es-san'-dria and Ma-ren'-go's known,
The last, where fought the great Napoleon.

Antwerp is noted for its cathedral, the spire of which is 441 feet high.

Ghent is the place where peace between the United States and Great Britain was concluded.

Waterloo is famous for one of the greatest battles ever fought; a battle that decided the fate of Europe and Napoleon.

The following extracts are from Byron's Waterloo:

And there was mounting in hot haste, the steed,
The mustering squadron, and the clattering car
Went pouring forward with impetuous speed,
And swiftly forming in the ranks of war.

Last noon beheld them full of lusty life,
Last eve in beauty's circle proudly gay,
The midnight brought the signal sound of strife,—
The morn, the marshaling in arms,—the day
Battle's magnificently stern array!
The thunder-clouds close o'er it, which when rent,
The earth is covered thick with other clay,
Which her own clay shall cover, heaped and pent,
Rider and horse, friend, foe, in one red burial blent.

Denmark.

Denmark comprises the peninsula of Jutland, the duchies of Holstein and Lauenberg, together with Fuen and Zealand, with the foreign possessions of Greenland, Iceland, Faroe Islands, &c.

The soil is fertile and well adapted to pasturage.

The atmosphere is thick and cloudy, but generally salubrious and healthy.

The Danes are honest and well educated.

The principal source of influence of this state, is the command of the entrance to the Baltic. It exacts a toll of all ships that pass in and out of that sea.

COPENHAGEN, on the island of Zealand, is the capital.

Holland.

The land of the Dutch, formerly called the Netherlands.

This is a flat, level country, below the bed of the sea; which is kept from encroaching and overflowing the land by means of dykes or embankments.

Canals serve the purpose of streets, and are the highways for the commerce of the kingdom.

The Dutch are the most inveterate smokers in the world.

They have colonies in South America, West Africa, Java, and other Asiatic islands.

AMSTERDAM, the capital, on Zuyder Zee, is noted for its canals, that serve the purpose of streets.

Italy.

A peninsula in the southern part of Europe; noted as having been the seat of the Roman Empire, and of the Popes, and as the land of sculpture, painting, architecture and music.

It is distinguished likewise for its mild climate, and as being the most delightful country of Europe.

It is now divided into several different states or governments, the principal of which are as follows:

1. The kingdom of Naples, or the two Sicilies. 2. The States of the Church. 3. Grand Duchy of Tuscany. 4. The kingdom of Sardinia. 5. The kingdom of Lombardy and Venice.

Upon Sardinia's isle, behold Sas-sa'-ri
High to the north, while south is one Cagl-ia'-ri.

Lombardy and Venice.

For her Cathedral known, the fair Mi-lan,
Upon the west of Lombardy we scan;
Venice, beside the *A-dri-at-ic* smiles,
High to the head on seventy-two small isles.
As Virgil's birth place, next Man-tu'-a know,
That keeps her station on the rolling *Po*.
Lo'-di is west of this, a warlike town,
Where Bonaparte a splendid victory won.

States of the Church.

Rome, by the *Tiber*, keeps her ancient seat,
Known for her temples and her structures great;
Her columns, arches, monuments we hail,
But the far famed St. Peter's first of all.
As fair Bo-logn-a keeps the northern border,
An-co'-na, to the south-east, boasts her harbor.

The Two Sicilies.

Naples.

Naples, near Mount Vesuvius, has long been noted for the beauty of its bay, the deliciousness of its climate, and the picturesque scenery in its vicinity.

Near Mount Vesuvius let Naples stay,
Long noted for the beauty of her Bay.
Pa-ler'-mo sits on Sicily's fair isle,
And there Mes-si'-na and Ca-ta'-ni-a smile,
As Syracuse is known for ancient splendor,
The wine cup to Mar-sa'-la we may tender.

Sardinia.

The kingdom of Sardinia embraces the island of Sardinia, and the northwestern part of Italy.

The latter has a fine soil and mild climate.

The Island of Sardinia has an area of about 10,000 square miles; it is a trifle smaller than Sicily.

A large portion of the surface is hilly and mountainous. It produces every variety of fruits common to southern Europe.

TURIN, the capital of Sardinia, on the Po, is noted for silks.

Genoa is noted as the birthplace of Columbus.

Marengo for a great victory of Napoleon over the Austrians, in 1800.

Lombardy and Venice.

Lombardy and Venice, or Austrian Italy, is situated between the River Po and the Alps.

It is one of the best cultivated states of Italy, and belongs to Austria.

Lombardy is in the west and Venice in the east.

MILAN, in the west of Lombardy, is noted for its cathedral.

Venice is situated on seventy-two small islands, at the head of the Adriatic, or Gulf of Venice.

Mantua is noted as the birthplace of Virgil.

Lodi for one of Napoleon's most splendid victories.

States of the Church.

Rome, the most celebrated city on the globe, is fifteen miles from the mouth of the Tiber. Among the structures and monuments of greatness, that excite the interest of travelers, is the great St. Peter's, the largest cathedral ever built.

Naples.

Naples, or the two Sicilies, includes the southern part of Italy, the island of Sicily, and the Lipari Isles.

These enjoy a warm climate, and have a fertile soil, that produces the greatest variety of grains and fruits.

The island of Sicily was formerly called the Granary of Italy.

Agriculture and manufactures, notwithstanding, are in a very backward state.

NAPLES, seven miles from Mount Vesuvius, is noted for its beautiful bay.

Syracuse is noted for its ancient splendor.

Marsala, for its delicious wines.

Grand Duchy of Tuscany.

One of the most populous states of Italy.

The people are industrious and enterprising.

Manufactures, as well as agriculture are flourishing.

FLORENCE, on the river Arno, is noted for painting and sculpture.

Pisa, for its leaning tower, 190 feet high, and 14 from a perpendicular.

Grand Duchy of Tuscany.

Florence.

Florence, situated on the Arno, is noted for its Gallery of Paintings and Sculpture. It is one of the most beautiful cities in Europe.

Flor'-ence is seated on the *Arno's* banks ;
In sculpture and in painting, high she ranks.
Pi-sa is noted for her leaning tower ; (pe-za)
Leg-born', near by, in commerce boasts her power.

Small States.

Mo-de'-na rules the Duchy of Modena ;
Carrara there, for marble 's known to many.
Ajaccio blooms on Corsica's rich coast,
And as Napoleon's birth-place let it boast.

Modena is the capital of Modena.

Carrara is noted for its beautiful marble.

Ajaccio, on the island of Corsica, is noted as the birthplace of Napoleon.

Turkey in Europe.

Near Bos'-pho-rus, in sight of Asia's shore,
Constantinople hears Mar-mo'-ra's roar ;
Well fortified, her harbor let her boast,
Her mosques and temples, but seraglio most.
And A-dri-an-o'-ple sits in ancient Thrace,
Upon *Ma-ris-sa's* banks her trading place.

Greece.

Ath'-ens, in Greece, a town of age and fame,
Beside the *Gulf E-gi'-na* writes her name.
Hy'-dra, upon an island takes her rest ;
Na-po'-li has a harbor called the best :
And Navarino is remembered yet,
For the destruction of the Turkish fleet,
In 18–27, when Frank and Russ, (1827)
And Britain, all combined the Turk to crush.
And Mis-so-lon'-ghi, last of all, beside
The Gulf Pa-tras', — the place where Byron died.

Turkey in Europe.

The seat of the Ottoman Empire, and the most southeastern country of Europe.

It is interspersed with mountains, valleys, and rivers.

It is mostly watered by the Danube and branches.

It has a fertile soil, and mild climate ; and under a liberal government would be one of the finest countries of Europe.

The court is called the Ottoman Porte, or Sublime Porte.

Constantinople, in the southeastern part of Europe, upon the Bosphorus, near Asia and the sea of Marmora, is one of the finest cities of Europe. The seraglio, or imperial palace, is a city of itself.

Greece.

One of the most distinguished countries on the globe. Noted for its great antiquity, and for having been the cradle of the arts and sciences.

It has lately been rescued from Turkish thraldom, by the combined aid of Russia, France and England ; and is now in a prosperous condition.

Athens is noted for its antiquity, and importance in former times.

Navarino is noted for the destruction of the Turkish fleet, October 20th, 1827, by the combined fleets of the French, English and Russians, under Admiral Codrington.

Missolonghi is noted as being the place of Byron's death ; April 19th, 1824.

Athens.

Athens, the capital of Greece, was one of the most celebrated cities of antiquity. It was anciently the seat of learning, eloquence, philosophy, poetry, and the fine arts.

Town of Syra.

Syra, on the island of Syra, in the Grecian Archipelago, is built on a conical hill, and has a very singular appearance from a distance. It is one of the most flourishing towns in Greece.

ASIA.

RUINS OF PALMYRA.—The ruins of Palmyra are among the most remarkable anywhere found, consisting mostly of temples, palaces, porticos, &c., of Grecian architecture. It is supposed to be the Tadmor in the desert, built by Solomon.

Asia is noted as the largest of the grand divisions of the globe; as having been the abode of our first parents, and the theater where most of the events, recorded in the scriptures, transpired, and as containing more inhabitants than all the rest of world.

Capes.

Cape *Tay'-mour's* parallel is *seventy-seven*, (77°)
Where one eternal winter finds a haven.
Near *seventy-two Svi-a'-toi's* limits run; (72°)
As *Skol'-at-skoi* is scarcely *seventy-one.* (71°)
East Cape, beneath the *Arctic Circle* lies,
While south and west, *Fa-de'-va* 's seen to rise;
And *Pack-a-chin'-skoi*, with *Lo-pat'-ka* lower,
From cold Kam-schat'-ka's bleak and frozen shore.
The Gulf Si-am' around Cambodia raves;
Ma-lac'-ca rears *Ro-ma'-nia* o'er the waves.
And *Cape Ne-gras'* lies westward from Ran-goon',
Where the Bay of Bengal heaves the watery moan.
South of Hindoostan, *Com-o-rin'* we see,
North eight, and east the seventy-eighth degree.
And *Mus-sen-doon'*, and *Ros-al-gad'*, belong
In Ar'-a-by, and finish out the song.

Rivers.

To-bol' and *Is'-sim* with the *Ir'-tish* join
In *O'-bi's* stream, that rolls to Obi's brine.
And *Yen-e-se'-i* with *Ton-goos'-ka* glide
Where *Pi-a-ci'-na* scours the Arctic tide.
There *An-a-bar'-a* and *O-lensk'* unfold,
With *Le'-na* in the same bleak ocean roll'd.
And *Ya'-na* drives, amid the frozen spray,
With *In-di-gri'-ca* and dark *Kal'-a-ma*.
And *An-a-dir'* pours eastward to the sea,
Last in the list of cold Siberia.

RIVERS OF THE EASTERN COAST.

Son-ga-ri drives, joined with the bold *A-mour'*,
In Tartary's Channel, from the Tartar shore.
Ho-ang' Ho and *Ki-ang Ku* eastward stray,
From Chinese shores into the Yellow Sea.
From China, to her sea, *Ho-ang' Ki-ang'*
For ages past, has 'neath the tropic sang.
O'er India's plains, China and Thibet too,
To China Sea, *Cam-bo-di-a* murmurs low.

RIVERS OF THE SOUTHERN COAST.

From Thibet southward rolls the *Ir-ra-wad'-da*,
O'er India's plains, a bold, gigantic body.
And *Bur-am-poo'-ter*, called by some *San-poo'*,
From Thibet comes, passing Hindoostan through.
Jum-na' and *Gan'-ges*, both of Hindoo birth;
Go-dave'-ry too, and *Krist'-na*, in her mirth,
To the Bay of Bengal send their billows forth.
Westward *Ner-bud'-dah* rolls to Cambay Gulf,
Smiling on Hindoo shores, as smiles the sylph.
The *In'-dus* empties in the A'rab Sea,
'Tween Hindoo, Af-ghan, and the Bel-oo-chee'.
Hel-mund' runs westward into Dura Lake,
Whose murmuring waves the Afghan borders shake.
The *Ted-zen* mingles with the Caspian brine,
'Tween Persian shores and Tartary the line.
And *Ji'-hon* rolls with *Si'-hon* by her side,
In Tartary, where Aral opens wide.
The *Ti-gris* and the great *Eu-phra-tes* joined,
Into the Persian Gulf with murmurs wind.

TOWNS AND COUNTRIES.

Siberia.

Traveling in Siberia.

The traveling in Siberia is performed mostly by means of dogs. Three, five, seven or more, as the load requires, are harnessed together before the light sleds, which are easily drawn over the ice and snow.

To-bolsk', upon the *To'-bol*, is the place,
Or chief abode of Russia's exiled race.
Ir-koutsk', on the *An-ga'-ri-a* river seen,
Chief town of East Si-be-ri-a, I ween.
Ki-ach'-ta on *Se-lin'-ga's* banks is laid,
The only spot where Russ and Chinese trade.
Ya-kutsk' on *Le-na*, Ok'-hotsk near the tide,
Are by the fur-trade, in one bond allied.

Japan.

On *Niph'-on Isle*, Jed-do stands first in place,
With near two millions of the human race.
Me-a'-co where *Diari* makes abode, (da-ee-'ree)
One hundred sixty from the Jeddo road.
And Nan-ga-sack'-i is the only port,
Where European traders make resort.

Independent Tartary.

And Bok'-ha-ra and Sam-ar-cand' abide,
In Tartary along the *Ko-huk* tide.
Ot'-rar and Tas'-cant by the *Si'-hon* keep,
By *Ji'-hon's* waters Balkh and Kie'-va sleep.

Siberia.

Siberia, or Russian Asia, is noted as being an almost unbounded expanse of level, frozen desert.

It extends from the Ural Mountains to the Pacific Ocean.

Some of the southern districts are fertile.

The Ural and Atlay Mountains yield gold, silver, platina and precious stones.

TOBOLSK, the chief town in Siberia, is noted as the residence of distinguished exiles.

Kiachta is the only place where the Chinese allow the Russians to trade.

Yakutsk and Okhotsk are the chief emporiums of the fur trade.

Japan.

A small empire east of Asia; comprising the islands of Niphon, Jesso, Kiusiu and Sikoke.

The inhabitants are the most civilized, the best educated, and sustain the best morals of any country of Asia.

It is the only country of Asia where the *rights of women*, are respected.

Their laws are very severe; quartering the body, immersion in hot oil, crucifying, &c., are among their modes of punishment.

The parent suffers for the child's crimes, and the child for the parent's.

JEDDO is one of the most populous cities on the globe.

Meaco is the residence of the Diari, or spiritual emperor, the head of the Sinto Religion.

Independent Tartary.

Noted for the independent and roving character of its inhabitants; and for its having been the seat of rule for Ghenghis Khan, Tamerlane and others.

It is bounded on three sides by mountains and deserts; and on the fourth by the Caspian sea.

The Tartars are subject to no foreign power, and are not united under any one government. They are a pastoral people. Their favorite food is *horseflesh*.

China.

Pekin.

Pekin, with the exception of London, is the most populous city on the globe. It is near the great Chinese Wall. It contains the palace of the Emperor, which forms one of its principal features. It is divided into the Chinese and Tartar city.

Pe'-kin, the first in population, stands
Near the great wall that guards the Chinese lands.
Nan-kin, known for her tower, from Pekin south,
Two hundred forty from *Ki-an'-ku's* mouth.

Canton, the great commercial town of China.

Canton is one of the most commercial cities of the Chinese Empire. Nearly all the teas sold to foreign nations are shipped from this port It was until quite late the only port Europeans were permitted to trade at.

In commerce, first, Canton, on *Canton River*,
Where Europeans sail, their *teas* to gather.

Chinese Tartary.

Yor'-kund by *Yorkund River*, finds a home;
'Tis Central Asia's chief emporium.
Maim-at'-chin makes the mountain pass her bed,
Where, with Ki-ach'-ta China holds a trade.

Yorkund is the emporium for central Asia.

Maimatchin, by a mountain pass, on the opposite side of the Sayanskoi Mountains, from Kiachta in Siberia, is noted as the only place at which the Russians are allowed to trade.

Chinese Empire.

Noted for its great antiquity, and for being the most populous empire on the globe.

It embraces China, Chinese Tartary, Corea and Thibet; the population of which is estimated at 250,000,000, the greatest number ruled by any one man.

The ruler of this immense mass of beings, is an absolute despot, but governs his subjects in a parental manner. He is styled the Son of Heaven. He belongs to the Mantchoo race, by whom China was conquered in 1644.

China.

The basis of the Chinese Empire; noted for the jealous character of its inhabitants, and for the *Tea plant*, which is cultivated to such an extent that it supplies the whole world. It is estimated that 60,000,000 pounds are annually exported to the United States and Great Britain.

The Imperial Canal is 600 miles in length.

The Great Wall is 1,500 miles in length, and twenty-four feet high. It is the greatest work ever performed by man.

Pressing the feet of females while children, to prevent their growth, is a prevailing custom among the Chinese, and is their criterion of female beauty.

The principal food of the nation is rice, though rats, puppies, mice, &c., are common dishes.

The army amounts to 800,000 men, but their mode of warfare is vastly inferior to the European.

Learning is highly prized in China, and is the only requisite qualification for office.

Nankin is celebrated for its porcelain tower, 200 feet high.

The Great Chinese Wall.

The Chinese Wall is unquestionably one of the greatest works ever performed by man. It was built by the Chinese, as a defense against the Tartars. It is 1,500 miles long, twenty-four feet high, and broad enough for several horsemen to ride abreast.

Chinese Tartary.

An elevated country on the table lands of the Himmaleh, Kuenlin and Celestial Mountains. It is a cold country, inhabited by a pastoral people, of whom but little is known.

Thibet.

Worship of the Grand Lama.

The worship of the Grand Lama constitutes the religion of a great portion of Central Asia. He is considered by his worshippers the Everlasting Father of Heaven. They believe that when he dies his soul passes into the body of some child, who is sought after by the priest, and immediately exalted to the throne.

Las'-sa, in Thibet makes her proud abode,
Where the Grand Lama sits, a human god.

Turkey.—Syria and Palestine.

A-lep'-po by an earthquake torn of late,
Is first in rule o'er little and o'er great.
On *Pharphar's* tide, Damascus makes her throne,
For silks call'd *damask*, and for sword blades known.
Jerusalem reigns just thirty miles from sea,
Jaf'-fa, her port, northwest, is known to be.
Southwest of all, Ga-za is on the coast,
For caravans it is a resting post.
Acre, from Jaffa north, her fortress rears;
On *Leb'-a-non*, one Diar-el Kai-mer peers;
Bal'-bec is by the feet of *Leb-a-non;*
Pal-my-ra in the *desert* lives alone.
Both these are known for relics of the past,
Where ruins rise on every side aghast.

Asia Minor

Smyrna is seated where *Le-vant'* is found,
As south the *Black Sea* dwells fair Treb-i-zond'.
Bru'-sa, near by the sea that's called *Mar-mo'-ra*,
Was once the capital of Turkish glory.
An-go-ra in the *interior* is built,
And famous for a goat with hair like silk.

Mesopotamia and Armenia.

Bas-so-ra is a place of wealth and trade,
On *Shut'-el A'-rab* is her station made.
Bagdad, that lives upon the *Tigris* shore,
Was once the seat of Calif rule and power.

Thibet.

Noted for being the most elevated country on the globe, and for the worship of the Grand Lama.

It is situated on the table lands of the Himmaleh Mountains, so elevated that the cold, in the winter season, is intense. The sky at all seasons appears as black as ink. The stars shine with the effulgence of suns; there is no twilight that precedes the rising, or succeeds the setting of sun or moon; and were there not mountain peaks of a still higher elevation, to foretell the opening or closing of day, it would be one sudden change from darkness to light, and from light to darkness.

Turkey in Asia.

Noted for its fine climate and fertile soil, and as having been the seat of most of the events narrated in Bible History, and the theater of more changes than any other part of the globe.

It comprises Syria, Palestine, Asia Minor, Mesopotamia and Armenia.

Syria and Palestine.

SYRIA.—Noted for its importance in former times, when Tyre, Damascus, Antioch, Balbec and Palmyra, were in their glory.

It was conquered by the Pacha of Egypt in 1832; but restored by the interference of the European powers in 1840.

PALESTINE.—Noted as the Holy Land, the inheritance of the Israelites, and as the theater where the most important events have occurred that the world has ever witnessed.

Asia Minor.

The peninsula between the Mediterranean and Black seas. Noted for its delightful climate, and as having been the seat of the kingdoms of Lydia and Troy.

Mesopotamia and Armenia.

MESOPOTAMIA was once the seat of the mighty Babylon; of Paradise; of the Tower of Babel; of the kingdom of Nimrod, Cyrus, Darius, Alexander, &c.

ARMENIA, north of Mesopotamia, is noted as the place where the Ark rested after the flood.

Towns of Turkey in Asia.

ALEPPO, the capital, once a flourishing city, was destroyed by an earthquake in 1822.

Damascus is noted for sword blades, and a silk called *damask*.

Jerusalem is thirty miles in the interior. Jaffa is its port.

Gaza is a resting place for caravans, before crossing the desert to Egypt and Arabia.

Mo-sul′ likewise drinks from the *Tigris* flood,
For *muslins* known, where Nineveh once stood.
Hil-lah, beside *Euphrates* makes her throne,
Built on the site of mighty Babylon.
Ar-me-nia's capital is Er-ze-roum′;
Van on *Lake Van*, a fortress of renown.

Arabia.

Mec′-ca, where old Mahomet took his birth;
With fair Me-di′-na where he veils his earth.
Yem′-bo, Medina's port, is by the sea;
Jid′-da is Mecca's port, all will agree.

Mocha.

Mocha, though in somewhat of a decayed state, is still the most important port of Arabia on the Red Sea. It is noted for its excellent coffee, which is carried to most parts of the world.

Mo′-cha, chief seaport town of Ar′-a-by,
Whose coffee's drank on every shore and sea;
Mus′-cat, a seaport town, well fortified.
The Sacerdotal prince, or Imam's pride.
Southeast it stands where the Persian Gulf unfurls,
And much renowned for trade in shells and pearls.

Persia.

Te-he′-ran, where the El′-burg peaks arise,
Heaves up her warlike forehead to the skies.
And Is-pa-han′, once capital, is lain,
In the *interior*, on a fertile plain.
Shi-raz′, famed for her wine and Persian lore,
Near where Per-sep′-o-lis in ruins lower.

Acre is noted for its strong fortress.

Balbec at the foot of Lebanon, and Palmyra, in the desert east of Balbec, are noted for their remarkable ruins.

Bassora, on Shut′ el Arab, is a place of great wealth and importance.

Bagdad was the seat of the caliphs.

Mosul, on the Tigris, near the ruins of ancient Nineveh, is noted for muslins.

Hillah, on the Euphrates, is supposed to be near the site of ancient Babylon.

Erzeroum is the capital of Armenia.

Van, on Lake Van, has a strong fortress.

Arabia.

The birthplace of Mahomet. It lies between the Red Sea and Persian Gulf. Noted for the unchanging character of its inhabitants; for its great antiquity; for its burning sandy deserts; its superior coffee, and as being the center of the Mahommedan Religion.

It is divided into Arabia Petræa, or stony Arabia, in the northwestern part; Arabia Deserta, or Desert Arabia, in the interior; and Arabia Felix, or Happy Arabia, in the southern part.

The Bedouins, [bed-oo-weens] or the wandering Arabs that inhabit the desert, subsist chiefly by robbery and plunder.

Mecca, the birthplace of Mahomet, is regarded as the capital.

Medina is important as the place of his tomb.

Muscat, the capital of Oman, and governed by the Imam, or sacerdotal prince, is the entrepot for the merchandise of the Persian Gulf, on which it is situated. It is noted for its extensive trade in pearls.

Moore's inimitable song of Araby's Daughter, has, among its other merits, the glow of oriental scenery.

Farewell—farewell to thee, Araby's daughter!
(Thus warbled a Peri beneath the dark sea:)
No pearl ever lay, under Oman's green water,
More pure in its shell, than thy spirit in thee.

But long upon Araby's green sunny highlands,
Shall maids and their lovers remember the doom
Of her, who lies sleeping among the Pearl Islands,
With nought but the sea-star to light up her tomb.

We'll dive where the gardens of coral lie darkling,
And plant all the rosiest stems at thy head;
We'll seek where the sands of the Caspian are sparkling,
And gather their gold to strew over thy bed.

Persia.

Noted for its great antiquity and importance in early times.

A large portion is barren, mountainous and destitute of running streams. It is the most fertile on the borders of the Caspian sea.

The Persians are the most learned of the Asiatic nations.

They manufacture the most beautiful carpets, silk shawls, porcelain, &c., in the world.

Bu-shire′, chief seaport on the *Persian Gulf;*
Or-mus, known once for commerce and for wealth.
Gam-broon′ and Min-ab, near to *Ormus* keep,
On the same shore, beside the coral deep.
Yezd, where the *Ghe′-ber* finds a last repose,
South of the desert blooms, as blooms the rose.
Sul-ta′-nia, found on I′-rack's northern beat,
The king's resort in summer from the heat.
Still farther north, in A-der-bi′-jan peers
Ta-breez′, a splendid town in former years.
Gour-gaun′, a fortress by the Tartar line,
On Persian shores east from the Caspian brine.

Afghanistan.

Cabul, on *Ka′-ma* tide, the Af-ghans greet,
Above the sea it stands six thousand feet.
O'er Ca′-bul's kingdom once Pesh-awer′ reigned
The first in rule, e'er Cabul was enchained.
And Can-da-har′ is by the dark *Hel′-mund,*
The central point where Door-au-nees′ abound.
Northwest of all, He-rat′, with Persia trades,
Where *Hin′-doo Koosh′* unfold their giant shades.

Beloochistan.

Ke-lat′ by *Mas′-kid River,* finds a seat,
On Mountains o'er the sea eight thousand feet.

Hindoostan.

Calcutta.

Calcutta, on the Hoogly, *an arm* of the Ganges, one hundred miles from the sea, is one of the most important cities of Hindoostan. That part of the city where the Europeans reside is magnificently built. Its commerce is very extensive; and the population is estimated at 625,000.

Calcutta, Hindoo's proud emporium, smiles
On *Ganges,* from its mouth one hundred miles.
On the same tide Ben-a′-res has a share,
Four sixty, northwest of Calcutta's lair. (460)

The inhabitants are well formed, and like the Dutch, are great smokers.

Teheran, strongly fortified, is at the foot of the Elberg Mountains.

Ispahan, once the capital, is on a fertile plain in the interior.

Shiraz, the seat of literature, and noted for delicious wines, is near the ruins of ancient Persipolis.

Yedz, near the center of Persia, is the resort of the persecuted Ghebers, or fire worshippers.

Sultania, in the province of Irack, is the summer resort of the sovereigns.

Tabreez was once a city of importance.

Gourgaun, east of the Caspian, and near the line of Independent Tartary, is a strong fortress.

Afghanistan.

The country which lies between Persia and Hindoostan.

The Afghans are a bold and warlike race; hospitable to strangers, and even to their most bitter enemies.

Cabul, on Kama River, is elevated 6,000 feet among the Hindoo Koosh Mountains.

Peshawer was once the capital of Cabul.

Candahar, on Helmund River, is the principal town of the Dooraunees.

Herat, in the northwestern part of Afghanistan, is the seat of trade between Persia and India.

Beloochistan.

The country lying south of Afghanistan. It is inhabited by a number of independent tribes, of whom the Beloochees are the principal. Like the Bedouin Arabs, they are a mixture of hospitality and ferocity; generous and liberal when hailed in their tents, but blood-thirsty and clandestine on the field.

Kelat, by Maskid River, is among the mountains, 8,000 feet above the level of the sea.

Hindoostan.

A large peninsula in the southern part of Asia. Noted for its great fertility, its peculiar religion, the superstitious character of its inhabitants, and for its great antiquity.

The Hindoo has the skin of the Negro, with the features of the European. They are extremely superstitious; servile to superiors, cruel to their women and inferiors, and destitute of moral honesty.

Their food is principally rice, which is raised to a great extent.

The cotton manufactures of this country have long been celebrated.

Benares, 460 miles northwest of Calcutta, on the same river, is one of the most populous cities of India, and noted as the seat of learning, and as a sacred city; thousands coming from various parts of Asia to end their days within its environs, regarding it as the sure gate to paradise. It is a great mart for diamonds.

Of gems and diamonds read her story o'er,
Of pilgrims dying, and of Bramin lore.
Pat'-na is on the *Ganges*, none can beat her,
Or match her for her *opium* and *saltpetre*.
Del'-hi, once capital of Hindoo rule,
On *Jumna* branch, is known to every school.
Cash-mere', whose shawls are of the Thibet goat,
Stands north of all, a city of much note.
La-hore' from Cashmere south, o'er Pun-jab shrouds
With Am-rit-sir', beneath her sunny clouds. (seer)
Su-rat', Bom-bay', Go -a, and Man-ga-lore,
Are found upon Hindoostan's *western shore*.
While south and east, Ma-dras' and Pon-di-cher'-ry
Along the *Cor-o-man'-del coast* may tarry.
Nagpoor' in the *interior* writes her name,
Where Hy-dra-bad' 'mid sparkling diamonds flame.

Farther India.

As *Ir-ra-wad'-da* rolls her billows south,
A'va is found five hundred up her mouth.
As Um-me-ra-poo'-ra north of this is seen,
Pe-gu' is on the *Delta* of the stream.
South of Pe-gu', where trade and commerce bloom,
On the *same tide*, behold the fair Ran-goon',
Ban -kok is o'er Siam a town of note,
On bamboo rafts one half the houses float.
Cam-bo'-dia's capital is called Sai-gon';
Beside *Cambodia's* mouth she takes her throne.
Hue, o'er Co'-chin China, next is seen, (oo-a')
Well fortified, and near the *Gulf Ton-quin'*.
On the peninsula's southern coast or shore,
Malacca reigns, with one called Sin-ga-pore'.

Further India comprises a territory of about 900,000 square miles, and has a population of 20,000,000.

The elephant here attains his greatest size, and is found in large numbers. The white elephant is highly valued, and in Siam and Birmah is an object of religious worship.

The Siamese are described as destitute of courage and moral honesty; and as being lazy and sluggish in their habits. They are puffed up with a national pride, and consider it a great disgrace to be in the employ of an European.

The governments of all these states are absolute despotisms. The throne and person of the sovereign is approached with the profoundest awe by the nobles and officers of state, who prostrate themselves before him, with their faces to the earth.

Females are not restricted here to the rigid customs of most Asiatic countries. Their faces are not veiled, or their company excluded from the other sex.

Patna is noted for its saltpetre and opium.
Cashmere is noted for its shawls.
Hydrabad, or Golconda, is noted for diamonds.

Farther India.

A large peninsula south of Thibet. Noted for its large, numerous, and majestic rivers, and for great fertility.

It comprises the Empire of Birmah and Assam; the kingdom of Siam and the British possessions.

AVA, the capital of Birmah, is on the Irrawadda.
Pegu is on the Delta of the Irrawadda.
BANKOK is the capital of Siam. It is noted for its floating houses, built on bamboo rafts.
HUE, capital of Cochin China, is a fortified town near the Gulf of Tonquin.

Vale of Cashmere.

Cashmere is a beautiful vale of the Himmaleh Mountains, in the northern part of Hindoostan. It is elevated 8,000 feet above the level of the sea; and enjoys a climate unequaled in mildness, save by the "Evergreen Quito," which it resembles. It was not long since in the possession of the Afghans; from whose rule it passed to that of Runjeet Sing.

The beauties of the Vale of Cashmere, are portrayed in the following graphic and glowing lines from Moore's Lalla Rookh:

Who has not heard of the Vale of Cashmere,
With its roses, the brightest that earth ever gave,
Its temples and grottoes, and fountains as clear
As the love-lighted eyes that hang over their wave?

Oh! to see it at sunset,—when warm o'er the Lake
Its splendor at parting a summer eve throws,
Like a bride full of blushes, when lingering to take
A last look at her mirror at night ere she goes!—
When the shrines through the foliage are gleaming half shown,
And each hallows the hour by some rites of its own.
Here the music of pray'r from a minaret swells,
Here the magian his urn full of perfume is swinging,
And here, at the altar, a zone of sweet bells
Round the waist of some fair Indian dancer is ringing.
Or see it by moonlight.—when mellowly shines
The light o'er its palaces, gardens and shrines;
When the water-falls gleam like a quick fall of stars,
And the nightingale's hymn from the Isle of Chenars
Is broken by laughs and light echoes of feet
From the cool, shining walks where the young people meet:—
Or at morn, when the magic of daylight awakes
A new wonder each minute, as slowly it breaks;
Hills, cupolas, fountains, call'd forth every one
Out of darkness, as they were just born of the Sun.
When the Spirit of Fragrance is up with the day,
From his Haram of night flowers stealing away;
And the wind, full of wantonness, woos, like a lover,
The young aspen trees, till they tremble all over.
When the East is as warm as the light of first hopes,
And Day, with its banner of radiance unfurl'd,
Shines in through the mountainous portal that opes,
Sublime, from that valley of bliss to the world.

PRONUNCIATION.

Indigrica,	*In-di-gre'-ca*	Thibet	*Tib'-et*
Balkh,	*Balk*	Pharphar,	*Far'-far*
Araby	*Ar'-a-be*	Chen Yang,	*Shen-Yang*
Caucassus,	*Cau-cash'-us*	Bakou,	*Ba-koo*

AFRICA.

SUEZ, a town of Egypt, on the southern part of the Isthmus, at the head of the Red Sea, and surrounded by a desert, is important as a caravan post between Egypt and Arabia; also for lying on the route of the British overland-mail, to Bombay.

Africa is noted for the dark complexion and degraded condition of its inhabitants; for its burning climate; its vast deserts, and its unknown and unexplored interior.

Capes.

Guar-daf'-ui Cape and *Orf'-ui*, part the tides; [orf-we
With *Bed-o-uin* and *Cape Bas-sa*, besides. [Bed-oo-win.
Then *Cape Delga'do*, east of Mo-zam-bique';
Of *Co-ri-entes'* let Mon-o-ma-ta'-pa speak.
Cape Am-bro's north of Madagascar Isle,
While to the south, *St. Mary* lives the while.
And south of all, *Good Hope* nods o'er the brine,
In thirty-five degrees below the *line*.
Si-er'-ra Fri'-o and the *Northwest Point*,
Are of Cim-be'-bas, as you're well acquaint.
Cape Le'-do, of Angola pass, and then
Coast Castle and *Three Points*, of Guinea, scan.
Pal-mas and *Mes-u-ra'-do* west of these,
In fair Liberia, smile above the seas.
Cape Verde, of Gambia; while *Sa-ha-ra's* shore
Has *Blan'-co*, *Bar'-bos*, and *Cape Ba-ja-dore'*.
Cape Spar'tel, near the Straits, and last in song,
And north of all, near Tunis throned is *Bon*.

PRONUNCIATION.

Guardafui,	*Gar-daf'-wee*	Orfui,	*Orf'-wee*
Bedouin,	*Bed-oo-ween'*	Bassa,	*Bas-saw'*

Rivers.

The Mediterranean sups the river *Nile*,
Whose waves o'er Nu'-bi-a and Egypt smile;
The *Sen-e-gal'*, the *Gam-bi-a* and the *Grande*,
Boil up from Senegambia's burning sand.
As *Mes-ur-a'-do* bids Liberia thrive,
In Guinea, *La'-gos* and *For-mo'-sa* live;
And *Ni-ger* here from Guinea rolls her tides,
And with *Ga-boon'*, in the Gulf of Guinea glides.
'Tween Congo and Lo-an'-go, *Congo* swells
From Ethiopia's scorched and unknown fields.
Co-an'-za's waves north of Ben-gue'-la course,
As on her southern limits roars the *Nourse*.
The *Orange*, from South Africa we track;
While in Cape Colony boils up the *Zack*.
Southward, the *Gau'-ritz* turns, bubbling forever;
As eastward of the Cape is *Great Fish River*.
Zam-bese' southeastward drives from Mo-zam-bique',
And bids her breakers the broad Channel seek.
Dark *Mu-ru-su'-ru* sleeps in Zan'-gue-bar,
Where farther north *O-zee'* provides a lair;
And last of all, from Abyssinia's shores,
In Bab-el-Man'-del Strait, fair *Ze-lia* roars.

TOWNS AND COUNTRIES.

Barbary States.

The Barbary States include Morocco, Algiers, Tunis, Tripoli and Barca ; or that portion of Africa north of the great desert of Sahara, and west of Egypt. It is distinguished for the number of its noxious animals; as the scorpion, serpents of a deadly venom, the hyena, the Numidian lion, and the destructive locust.

The people of these states were once extensively engaged in piracies. The present inhabitants are Moors, Jews, Arabs and Berbers.

Morocco.

Morocco, near Mount Atlas, holds her reign;
Unfolded on a smooth and fertile plain.
Fez, for her learning, once could boast with pride;
Southwest from Fez, is Me'-qui-nez espied.
The largest ports, Ba-bat' and Mogadore',
Are found along Morocco's western shore ;
Where European consuls take their fare,
Close by Gibraltar Straits, is found Tan-gier' ;
Known for her pirates once, behold Sal-lee !
That keeps her station by the roaring sea.

Algiers.

Oran and Bona in Algiers arise ;
The last, for coral fisheries, we prize.
There Con-stan-ti'-na smiles in antique mood,
And old Algiers boasts of her pirate blood.

Tunis.

Tunis southwest the Carthagenian throne,
In Tunis reigns superior and alone.
Kair-wan', from Tunis south, famed for her mosque,
Finds an abode upon the Barbary coast.

Tripoli.

And Trip'-o-li, in Tripoli we scan ;
Where from the interior comes the caravan.

Barca.

On Barca's northern shore, is seated Derne;
Cy-re'-ne's tombs with wonder there we learn.

Darfoor.

And in Dar-foor', Cob-be' as monarch reigns,
Where laughs Tam-bul', above her fertile plains.

Fezzan.

With Germa, o'er Fezzan' Mour-zouk' may shroud,
Mourzouk is compassed round by walls of mud.

Morocco.

In the northwestern part of Africa The Mauritania of the ancients; embracing Morocco, Fez, and Tanfilet.

The government is an absolute despotism. Agriculture is neglected, and the only manufacture is morocco leather, made of goat skins.

Morocco, the capital, is on a fertile plain, twelve miles from Mount Atlas.

Tangier is noted as the residence of most of the European Consuls.

Algiers.

The ancient Numidia; situated east of Morocco. It is the most fertile and healthy of the Barbary States. Noted for the coral fishery on its coast.

It was invaded and conquered in 1830, by France, and is now a part of the French dominions.

Algiers, once called the Pirate Nest, is built on a hill.

Tunis.

The ancient Africa Propria. Noted as the seat of ancient Carthage, so long the rival of Rome.

The government is more liberal, and the people more civilized than any of the other States.

Tunis is noted as being the capital, and as lying near the site of ancient Carthage.

Tripoli.

The ancient Tripolis; it is a dry, sparsely populated country; fertile on the coast, but mostly desert elsewhere.

It is the weakest of the Barbary States; but its inhabitants are among the most civilized.

It abounds in ruins and relics of past ages.

Barca.

The ancient Lybia ; it was once famed for its three crops a year, but is now mostly a desert. It once contained the temple of Jupiter Ammon.

Darfoor.

A large oasis in the southeastern part of Sahara. The inhabitants are Mahommedans. The government is a rank despotism.

Fezzan.

The largest oasis in the world. It is south of Tripoli, to which country it belongs.

Egypt.

Pyramids and Sphynx.

The Pyramids of Egypt are among the most remarkable works of antiquity. They are on the west bank of the river Nile, and about forty in number. The largest is five hundred feet high, and 728 feet at the base. The Sphynx is a monster cut out of the solid rock, having the head of a man and the body of a lion. It is one hundred and twenty five feet in length. It is now mostly buried in the sand.

Fair Cai-ro and Ro-set'-ta standing where
Egyptian ruins cloud the middle air:
There Thebes and Alexandria lie unfurled,
The dim resemblance of an ancient world.

Senna Gambia.

Temboo, St. Louis, Bathurst and Kem-i-noo',
In Senna Gambia stand in open view.
The first is capital, the chief of all,
The next, on Senegal, belongs to Gaul.

Sierra Leone.

In Sier'-ra Le'-one, Freetown let us write,
Reformed and christianized from heathen night.

Liberia.

Mon-ro'-via, in Liberia we see,
Where Afric's sons are numbered with the free.

Guinea.

In Guinea stand Bi-af'-ra and Be-nin',
There Ab'-o-mey—a pagan rude is seen.
Coo-mas'-sie, where Ashantee's tribes abide,
And push their conquests round on every side.

Loango and Congo.

Lo-an'-go, on Loango's coast unfolds,
And Con'-go's skies St. Salvador beholds.
The last is throned upon a mountain high,
And famed for health beneath a cloudless sky.

Egypt.

One of the most celebrated countries of antiquity, the cradle of the arts and sciences, the seat of the kingdom of the Pharaohs, is situated in the valley of the Nile, in the northeastern part of Africa.

It is now noted for its stupendous ruins, that attest its former greatness.

It is at present the seat of a new and prosperous kingdom, under Mahommed Ali, who has lately introduced European arts, learning and civilization into the kingdom.

Grand Cairo is the largest city of Africa, and is the residence of the Pacha of Egypt.

Rosetta, Thebes and Alexandria, are all noted for the remarkable ruins found in their vicinities.

Senna Gambia.

A well watered and productive country, south of the Great Desert.

The climate is hot and fatal to Europeans.

The English, French and Portuguese have settlements on the coast.

Temboo is the capital.

St. Louis is claimed by France.

Sierra Leone.

Established by Great Britain, 1787, for the purpose of Christianizing the natives.

The colony contains about 18,000 inhabitants; mostly negroes, taken from slave ships.

Freetown is a missionary station, established by Great Britain, 1785.

Liberia.

Formerly an American colony—now an independent republic. It was colonized in 1821; became independent in 1847.

Monrovia, the capital, was founded by the American Colonization Society, 1820.

Guinea.

Comprises the kingdoms of Ashantee, Dahomey, Benin, &c. Noted for its burning climate. The coast is divided into the Grain, Ivory, and Gold coast.

Coomassie, the largest town in Guinea, is the capital of Ashantee, the most powerful kingdom in the West of Africa.

Abomey is but a large collection of huts. Barbarism and paganism exist here in their most hideous shapes.

Loango, Congo, Angola, and Benguela.

Loango is about 400 miles in extent. The climate is said to be salubrious. The coast is high.

Congo is bounded on the west by the Atlantic; on the east by lofty mountains.

Angola is resorted to by slave vessels, to procure slaves from its coast.

Benguela.—The coast is extremely unhealthy.

South Africa.

Cape Town.

Cape Town, in Cape Colony, on the extreme southern shore of Africa, was founded by the Dutch in 1650, and is now in the possession of Great Britain. It is the great half-way-house for vessels in the China or India trade.

Cape Town, within Cape Colony is found,
Where vessels stop when to the Indies bound.
And from Cape Town, northeast, we likewise view
Kur-re-chan-ee′, Ma-show′, and Lat-ta-koo′.

Mozambique.

In Mo-zam-bique′, holds Mozambique her rule,
Which with So-fa′-la's owned by Portugal.
There Quil-li-mane′ and In-ham-bane′ behold
Where Lisbon trades for ivory, slaves and gold.

Zanguebar.

In Zan′-gue-bar, dark Mag-a-dox′-a breathes,
And there Me-lin′-da with Quil-lo′-a lives.

Adel.

And A′-del and Ber-be′-ra both appear,
Where Adel's plains their tawny bosoms rear:
For gums and frankincense, and costly myrrh,
These both are known and chronicled afar.

Abyssinia.

And Mas-sua, Gon′-dor, and one Ax′-um throng,
Where Abyssinia's doors are round them hung.
As Axum spreads her ruins to the day;
Gondor is on a hill, and built of clay.

Nubia.

And Sen′-na-ar, Shen′-di, and Mer-a′-weh smile
With Derr in Nubia, on the flowing *Nile*.
Meraweh's famed for temples, near her border,
Shendi for pyramids of ancient order.
As by the *Nile* Dongola mounts the throne;
Ip-sam′-bul for her temple well is known.

South Africa.

Comprises Cape Colony, Caffraria, the Land of the Hottentots, and the District of the Boshuanos.

Cape Colony was settled, in 1650, by the Dutch, and is now in the possession of Great Britain.

Caffraria, or the Country of the Caffres, extends about 650 miles along the eastern coast of South Africa.

The Caffres are a mixture of the Arab and Negro. They possess vigorous constitutions, have brown complexions, with features of an European cast.

The Bushmen, or Wild Hottentots, are among the most degraded of the human species. They have sharp, fierce-looking features, and a wild expression in their eyes. They wander about without any fixed habitation, subsisting on roots, toads, lizards, grasshoppers, &c.

Mozambique.

A large country, on the eastern coast of Africa, claimed by Portugal. Its trade is ivory, slaves and gold.

Mozambique, and all the rest of the ports on the coast of Mozambique, are in the possession of the Portuguese, who hold a traffic with the natives for ivory, gold and slaves.

Zanguebar.

A marshy, unhealthy country, that abounds in elephants, crocodiles and venomous serpents.

Adel and Berbera.

Adel, west of Berbera, is imperfectly known.

Berbera is the most eastern part of Africa, and noted for gums and spices.

Adel and Berbera, the chief towns, are noted for their frankincense and rich gums.

Abyssinia.

The ancient Ethiopia, is an uneven country, intersected by ranges of high mountains. The soil in the valleys is fertile, and the climate is mild and salubrious.

The inhabitants are a cruel and licentious race.

Axum is noted for monuments and ruins; among which are 40 obelisks; one 80 feet high.

Nubia.

A rocky, sandy, desert country, where pillars of sand are seen moving in the wind, and where the poisonous simoom blows. It belongs to the Pacha of Egypt.

Near Meraweh are a number of temples, adorned with sculptures, hieroglyphics, &c. One of these, the largest, is 450 feet in length and 160 in width.

Near Shendi are upward of 40 pyramids, supposed by many to be older than those of Egypt.

Dongola, on the Nile, is the capital.

Ipsambul is noted for a temple of immense proportion, excavated out of the solid rock. It is adorned with colossal statues and painted sculptures.

Central Africa.

Near *Niger's* banks Tim-buc′-too finds a spot,
For caravans a place of great resort.
Se-go′ and Jena both are towns of trade,
Southwest from this, beside the *Niger* laid.
From *Niger* east, some hundred miles or more,
Is Sack-a-too′, the first in size and power.

The commerce of Africa, Arabia, and many other parts of Asia, is carried on by means of caravans. The camel is the only animal that can endure the scorching heat of the sandy deserts. A caravan sometimes consists of 2000 camels, and as many persons.

Soodan, or Central Africa.

Soodan, or Nigritia, sometimes called Central Africa, is imperfectly known. It includes all south of the Great Desert, and north of Ethiopia.

TIMBUCTOO, once supposed to be a large city, is found to be but a mere collection of huts.

Great Desert.

The Great Desert of Sahara, north of Soodan, is 3000 miles long and more than 1000 broad, containing over 1,800,000 square miles. This immense expanse is nearly all covered with sand, which is blown by the wind in moving pillars, scattering death in its fearful path.

[The Red Sea, viewed from Ras Mahommed, on the southwest coast of Arabia Petræa.]

The RED SEA is a large inlet or bay, communicating with the Indian Ocean by the Strait of Babel-Mandel and the Gulf of Aden. Its length is about 1400 miles, and its greatest breadth 200. This sea is bounded on the east by Arabia; on the west by Egypt.

It is still memorable for the wonderful passage and safe deliverance of the children of Israel through its waters; and for the overthrow of the haughty Pharaoh and the Egyptian host.

The celebrated Song of Miriam, sung after this great drama (see *Exodus* xv, 20), is thus paraphrased by MOORE, in one of his most beautiful and melodious strains.

MIRIAM'S SONG.

SOUND the loud timbrel o'er Egypt's dark sea!
Jehovah has triumph'd,—his people are free
Sing—for the pride of the tyrant is broken,
His chariots, his horsemen, all splendid and brave—
How vain was their boasting!—The Lord hath but spoken,
And chariot and horseman are sunk in the wave.
Sound the loud timbrel o'er Egypt's dark sea!
Jehovah has triumph'd,—his people are free.

Praise to the Conqueror, praise to the Lord!
His word was our arrow, his breath was our sword!—
Who shall return to tell Egypt the story
Of those she sent forth in the hour of her pride?
For the Lord hath look'd out from his pillar of glory,
And all her brave thousands are dashed in the tide.
Sound the loud timbrel o'er Egypt's dark sea!
Jehovah has triumph'd,—his people are free.

ISLANDS.

Chained to the Arctic sea is Greenland found,
Where winter spreads his desolation round.
As Disco here in snowy garb is dressed,
Prince William's Land from Baffin's Bay is west;
Southampton keeps in Hudson's ample bay,
While west of all, Sabine and Melville lay.
And Newfoundland from Labrador is south,
Where the St. Lawrence river opes her mouth.
Prince Edwards here, with Anticosti keeps,
With one Cape Breton, on the liquid deeps.
Long Island floats upon the azure wave,
Where Martha's Vineyard and Nantucket lave.
Ber-mu-das and Ba-ha'-ma, blooming where
Sweet spring distils her ever-balmy air; [smiles,
Where storms and earthquakes frown, and verdure
In summer's climes lay fair West India Isles;
Of these rich Cuba sparkles o'er her stand;
Hay'-ti and Por-to Rico join the band; [torn,
Though scorched by lightnings, and by earthquakes
Ja-mai'-ca there still blushes like the morn.
Southeast from these, and smiling on the tide,
Ca-rib'-bee's mounds are mantled in their pride;
There Bar-ba-does' and fertile Gua-da-loupe',
With Trin-i-dad', stand in the elfin group;
Jo-an'-nes dwells in Am-a-zon's broad mouth,
With Mar-tin-Vas', and Sax-em-burg more south;
Au-ro-ra and South Georgia, dismal shores,
Where winter with his blustering tempests roars;
And Ter'ra del Fu-e'-go, scorched by fire,
With Falk'-land, 'neath the storm's impetuous ire;
South Shet'-land and South Ork'-ney, unexplored,
With Sand-wich Land, whose names we scarce afford;
And St. Hel-e'-na, where Napoleon lay,
Is on the western coast of Africa.
As-cen'-sion and St. Mat'-thew northly glow,
With one St. Thomas, and Fer-nan-do Po;
Cape Verd, from Gambia west, comes in the song,
As the Canary Isles to Spain belong;
Madeira there with sparkling wine cup full,
In mountain garb, is owned by Portugal;
For health renowned, then comes the fair Azores,
Or Western Isles, where ocean's dark surf roars.

GREENLAND is probably the largest island in the world, excepting New Holland; it is known to extend more than 1,400 miles north, and how much further is unknown. It probably reaches to, or beyond the pole, and forms an Arctic Continent of itself.

MELVILLE is noted as having been the head quarters of Captain Parry, for two years.

NEWFOUNDLAND is noted for the greatest codfisheries in the world. It belongs to Great Britain.

NANTUCKET is noted as a whaling depot.

LONG ISLAND, south of Connecticut, is noted for its fertility of soil.

THE BAHAMAS and BERMUDAS, are noted for their salubrious climate. St. Salvador, one of the Bahamas, was the first land discovered by Columbus.

THE WEST INDIES are noted for their great fertility.

CUBA, the largest, belongs to Spain; it is about 800 miles in length, with an average width of 75 miles.

JAMAICA, one of the most beautiful of the West Indies, is subject to hurricanes, earthquakes, and dreadful storms of thunder and lightning.

JOANNES is a large island, lying in the mouth of the Amazon.

TERRA DEL FUEGO, or the land of fire, is a cold, desolate region, inhabited by a race of the most miserable savages.

ST. HELENA is a rocky island off the coast of Africa. It is noted as having been the prison of Napoleon, from 1815, to his death, 1821. His body remained there till 1840, when it was taken to France.

ASCENSION is noted for turtles.

CAPE DE VERDES have a hot, unhealthy climate.

THE CANARIES are noted for canary birds, and the Peak of Teneriffe, an extinct volcano, 12,250 feet high.

THE AZORES or Western Islands, belonging to Portugal, are noted for fertility of soil and salubrity of climate.

MADEIRA, a mountainous island, is noted for fertility and Madeira wine.

THE HEBRIDES belong to Scotland, they are mostly barren and sterile.

THE SHETLANDS, north of the Orkneys, number in all about 100. They are cold and barren.

ICELAND, one of the largest islands in the world, is noted for Mount Hecla, and its geysers or springs of hot water. The climate is dreary and cold. It is owned by Denmark.

SPITZBERGEN is the most northern land known; it lies between the 77th and 81st degrees of north latitude. On its coast are found whales, sea-dogs, sea-cows, sea-lions, &c. In the summer the sun does not set for three months.

Great Britain, west of Europe, takes her post;
And Ireland borders on her western coast;
And Fa'-roe, Shet-land and the Ork'-neys gaze
Still further north, where sleep the Heb'-ri-des;
Fu-en' and Zeal-and east of Denmark keep;
Born-holm and *Ru'-gen* in the Baltic sleep;
O'-land and *Goth-land* there in slumbers lay,
And *A'-land* gazes o'er the Baltic sea.

Of the Mediterranean.

Mi-nor'-ca and *Ma-jor'-ca*, east of Spain,
With *Iv'-i-ca* assert their watery reign;
Sar-din'-i-a on her watery throne I found,
With *Cor'-si-ca*, her sister, by her crowned;
Cy'-press and *Can'-di-a* in angelic mien,
With *Sic'-i-ly* in the same bright sea are seen.

Of the Arctic Ocean.

The foxes' empire, *No-va Zem-bla*, stands,
And o'er the pole *Spitz-ber-gen* holds her hands.

Of the Indian and Pacific Oceans.

Com-o'-ro Isles, *Bour-bon*, and *Isle of France*,
With *Mad-a-gas'-car*, from the waves advance;
And *Lac'-a-dives* and *Mal-dives* there are strown,
With *Cha'-gos Isles*, by Indian zephyrs blown.
South of Hindoostan blooms the fair Cey-lon',
Known for her costly pearls and cinnamon;
Hai-nan' is seated in the gulf Tonquin, (ton-keen')
From China east, Formosa Isle is seen,
With Ki-u-si-u and So-koke, we scan
Niphon and Jesso, islands of Japan.
From Niphon north behold Saghalien Isle,
While north and east are those we call Kurile,
And Bor'-ne-o where the ourang-outang is found;
Whose shores with forests and with swamps abound,
And Cel'-e-bes, where herbs of poison grow,
And reptiles live, stands east of Bor'-ne-o;
Sumatra where Mount Ophir towers the while,
As Java slumbers a volcanic isle.
Moluccas for their spices next we name,
As the Philippine Isles are owned by Spain.
Australia, Ocean's first born offspring stands,
And o'er his azure empire spreads her lands,
New Guinea and New Zealand there are lain,
And there Van Dieman's Land usurps her reign.

Nova Zembla lies north of Europe and Asia. It is destitute of all traces of vegetation, save lichens and mosses. Yet on its shores are found vast numbers of foxes, white bears, walruses and seals.

Corsica, 100 miles long, and about 44 wide, is noted as the birthplace of Napoleon.

Sardinia, 160 miles long and 60 wide, is rich with minerals, and has a fertile soil.

Sicily was once called the granary of Europe. It is the largest island in the Mediterranean, and is noted for Mount Etna.

Madagascar, on the coast of Africa, is one of the largest islands in the world, being 840 miles long and 300 wide. Its inhabitants are Arabs, Negroes and Malays. The soil is rich and fertile, and the climate healthy.

Bourbon belongs to France. It contains a volcano in a state of activity.

The Isle of France, or Mauritias, belongs to England. It is noted for a lofty mountain, which is crowned by a high, rocky peak, called Peter Botte Mountain.

New Holland, or Australia, is the largest island in the world, having an area of 3,500,000 square miles. The whole of this vast tract of land is claimed by Great Britain.

The natives or aborigines of this island are probably the lowest in the scale of any that belong to the human family. They are the only race that goes entirely naked. Their food consists of fish, snakes, snails, worms, lizards and all kinds of loathsome reptiles.

Van Dieman's Land, situated south of Australia, is noted as being the place where most of the convicts of Great Britain are now banished. The population is about one third criminals.

New Zealand became a part of the British Empire in 1840. The natives are tall and well formed, and were formerly cannibals.

Sumatra is noted for Mount Ophir, 13,000 feet high. The island produces large quantities of camphor and pepper.

Java belongs to the Dutch. It produces coffee, sugar, rice, &c. Batavia, the capital, is a great commercial emporium for the trade of the Dutch in the East.

Celebes is noted for its vast number of venomous reptiles, flies, &c., that annoy the inhabitants to such a degree that they are compelled to build their houses on posts, to prevent their intrusion.

The Philippines are noted for terrific storms of thunder and lightning.

Borneo is one of the largest islands in the world. Its shores are beset with swamps and forests. The orang outang, the connecting link between man and the lower animals, is found here.

The original inhabitants of the Ladrones have been nearly all exterminated by the Spaniards.

The Caroline Islands are mostly all of coral formation. They are beat by a tempestuous ocean, and are subject to storms and hurricanes.

The Caroline, where reefs of coral form,
Brave the rough surf, the tempest and the storm ;
Ladrones beneath the Spanish yoke are bound,
While farther north the Bonin Isles are found ;
The Sandwich Islands, where Mount Roa keeps,
And where Kirauea flames above the deeps,
Where Captain Cook was by the natives slain,
Are bound together in the coral chain.
Folded in Ocean's arms, the Friendly Isles,
By the Society, rear up their piles ;
Fair Otaheite, in the last named band,
Shines like an Eden in a fairy land.
Marquesas Isles are in the burning zone,
South of the line with those called Washington.
Ju-an' Fer-nan'-dez sparkles in the deeps,
And young Chi-lo'-e near to Chili keeps ;—
As Gal-a-pa'-gos fronts the torrid skies,
Van-cou'-ver's to the north at fifty lies ; (50°)
And farther still, Queen Charlotte's Isle is sown,
Where On-a-las'-ka and A-leu-tian shone.

The Aleutian Islands, in the North Pacific, belong to Russia. They are about forty in number, and contain several active volcanoes. In 1795 a volcanic island rose from the sea, which, in 1807, had enlarged to twenty-one miles in circumference.

The natives of these islands are a mild race of savages, who live in large subterranean houses, which frequently contain from 100 to 150 persons.

THE SANDWICH ISLANDS are among the most important of the Pacific. The native inhabitants have all been converted to the Christian religion.

Otaheite.

Otaheite "the gem of the Pacific," is the largest of the Society Islands. Its circumference is about 108 miles. The interior rises into high mountains, the sides of which are covered with rich verdure. The natives of this island are tall and well made; they have lately been converted, by the efforts of missionaries, to the Christian religion.

Juan Fernandez.

Juan Fernandez was formerly noted for having been the solitary residence of Alexander Selkirk for several years; from which event sprung the celebrated romance of Robinson Crusoe. It has been described as one of the most beautiful islands in the world.

Ladies of the Azores, or Western Islands.

MOUNTAINS.

Mt. Chimborazo.

North America.

The *Rocky Mountains* join in airy bands,
O'er British soil and over Yankee lands.
O'er Mexico and Guatamala, too,
In the same chain, *Cor-dil'-le-ras* we view.
As *Ozark Mountains* in Missouri pile,
In Tennessee is *Cumberland* the while,
N. C., Virginia, Maryland and Penn.,
Are bound together by the *Blue Ridge* chain.
O'er the same states, except the state N. C.,
The *Alleghany* keeps them company.
The dark *Green Mountains* in Vermont embower,
And the *White Mountains* o'er New Hampshire tower.

South America.

O'er South America the *An'-des* rise,
With *Chim-bo-ra'-zo* throned above the skies.
So-ra-to, too, the highest peak, is there;
Bolivia is the place he makes his lair.

Europe.

As Scotia's climes the proud *Ben Ne'-vis* hails,
With *Grampian Hills*;—*Snowdon* is found in Wales.
With huge Cantabrian and Iberian reign
The bold Ne-va'-da o'er the realms of Spain.
Castile, To-le-do, and Mo-ra'-na steep,
O'er Spain and Portugal their sentries keep.
'Tween France and Spain behold the *Py-ren-nees'*;
The proud *Ce-vennes'* in France the traveler sees,
Au-vergne, near by, spreads out his rocky line;
As the Vosges Mounts are west the river Rhine;

The following table shows the length of the principal ranges of Mountains:

	MILES.
Andes,	4,500
Mexican and Rocky Mountains,	5,500
Whole American Chain,	10,000
Altain Mountains,	5,000
Mountains of the Moon,	2,000
Ural Mountains, Atlas Mountains,	1,500
Dofrafield Mountains,	1,000
Olonetz,	1,000
Alleghany,	900
Alps,	600
Appenines,	700
Carpathian,	500
Green Mountains,	350
Pyrennees,	200

The following shows the hight of some of the loftiest peaks of Mountains:

		FEET.
Chumularee,	Thibet,	29,000
Sorato,	Bolivia,	25,000
Chimborazo,	Equador,	21,440
Hindoo Koosh,	Afghanistan,	20,600
Cotopaxi, a volcano,	Equador,	18,890
St. Elias, highest in N. A.,	Russian Am.	17,900
Popocatapetl, highest in	Mexico,	17,700
Mt. Blanc, highest in Europe,	Italy,	15,685
Antisana Farm House,	Equador,	14,300
Mount Etna, volcano,	Sicily,	10,950
Mount Lebanon,	Syria,	10,000
Mount Sinai,	Arabia,	8,168
Pindus, highest in	Greece,	7,677

The highest inhabited spot in Europe, is the Monastery of St. Bernard, in the pass over the Great St. Bernard Mountain. It is 8,000 feet above the level of the sea. Here the monks entertain all strangers and travelers gratis, for three days. Dogs are so trained that they are sent out in the storms of snow, to rescue benighted travelers.

In South America we find large cities excelling the above. They are mostly on the table lands of the Andes. Quito is 9,000 feet above the level of the

As Switzerland claims the *Alps*—the *Ap'-pen-ines*
O'er Italy unfold their snowy shrines.
On Austrian shores, upon the map are traced
The *Erz'-ge-berg*, with the *Car-pa'-thi-an* braced.
He'-mus in Turkey, with the *O-lym'-pus* mound;
While proud *Par-nas-sus Mount* in Greece is found,
The *Dof-fra-field* in Norway, and between
Norway and Sweden, on the map are seen.
O-lentz' in Finland, while the *U'-ral* chain
'Tween Russia and Siberia may reign.

Asia.

In Turkey dwells *Tau'-rus* and *Lebanon;*
As *Ar-a-rat'* is there on his high throne.
Ho'-reb and *Si'-nai* in their grandeur tower,
With one *Ram-le-ah,* on the Arab shore.
Par-a-po-mi'-sus and the *Eldwin* brood,
With *Lou-ris-tan'* o'er Persia's neighborhood,
The *Gon-do-ree'* and *Kind,* with *Hindoo Koosh',*
O'er Afghanistan shores their shadows push.
Him-ma'-leh Mountains bound Hindoostan north;
Hindoostan is the place where *Ghaut* has birth.
From Thibet north, *Ku-en'-len Mountains* peer;
In Chinese Tartary the *Celestials* rear.
Al-tay', Sai-an'-skoi and the *Ya-blo-noy',*
Along Siberia south, we next espy.
Stan-voy' is east, near the Pacific coast,
Where O'-kotsk's billows round their feet are tossed.

Africa.

In Barbary the *Atlas Mounts* belong;
South of Nigritia is the chain called *Kong.*
Kong Mountains join the *Mountains* of the *Moon,*
In Ethiopia, 'neath the burning zone.
The *Cam-e-roon',* in Guinea next we see;
As the *Snow Mounts* are in Cape Colony.

sea; La Paz 12,000; Guanca Velica and Potosi reach as high as 13,000 or 14,000 feet; and the farm house of Antisana, the highest inhabited spot on the globe, is sublimely elevated at the hight of 14,300 feet.

The sublimest mountain scenery in any part of the world, is found in South America. The cities just enumerated are above the region of the clouds and storms, and enjoy one perpetual spring, with the clear azure above, which is lit by day with the great luminary, and by night sparkling with the effulgence of ten thousand stars.

Travelers in ascending the Andes have witnessed storms of lightning and thunder raging in their elemental fury, thousands of feet below them, while they themselves were enjoying the cool zephyr, or the mild sunshine.

The Andes, seen from the Pacific Ocean off the coast of South America, present one stupendous wall of adamant, that in the distance has a hazy, blue appearance, which contrasts and softens with the clear white of the eternal snow with which the top or summit is crowned.

The highest peak of the Andes is Mount Sorato, in Bolivia; its summit is elevated 25,000 feet. Illimani, near Sorato, is the second highest, being 24,350 feet. Chimborazo, in Equador, is the third in elevation, being 21,444 feet.

The Alps are the highest mountains of Europe, and among the Alps, Mount Blanc, (or the White Mountain), towers above all others, being 15,685 feet. It is in the northern part of Italy.

> Mount Blanc is the monarch of mountains,
> We crowned him long ago,
> On a throne of rocks, in a robe of clouds,
> And a diadem of snow.
> Around his waist is the forest braced,
> And the avalanche in his hand.
> But e'er it fall, the thundering ball,
> Must pause for my command.
>
> *Manfred.*

The highest peak of Asia is Chumularee, of the Himmaleh range, being 29,000 feet. This is the highest mountain in the world. Next to this is Dawalegeri, 27,677 feet. Over twenty different mountains in this chain are said to be over four miles in hight.

Mountains are sometimes intersected by rivers, which afford in many places but a narrow channel. The passages of the Potomac and Susquehannah, through the Blue Ridge, and the Missouri through the Rocky Mountains, are the most distinguished.

Mountains are great obstructions to roads and canals, as well as rivers. The roads over the Andes are so dangerous and difficult that they can be passed only by mules and lamas. They are often constructed upon the side of the mountain precipice, where a single misstep would precipitate the traveler thousands of feet into the yawning gulf, or chasm beneath. The pass of Quindu, between Popayan and Bogota, excels all others. The highest part of the road is 11,000 feet above the level of the sea. "No hut," says a distinguished writer, "is to be seen for eleven days; the path winds through chasms for half a mile in length, and such places are covered with the bones and carcasses of animals that have perished from fatigue or accident."

VOLCANOES.

View of Stromboli.

Stromboli, on the Lipari Islands, north of Sicily, is one of the most active volcanoes in the world. It has burned for more than two thousand years without interruption. It is visible at the distance of more than 100 miles, and is styled the great Light House of the Mediterranean

Mount St. Elias is a mount of flame,
Near the Pacific, in the Russian claim.
And *Po-po-cat-a-petl*, in Mexico,
Has a high summit covered o'er with snow;
In Guatimala, *Cos-a-gui'-na* piles,
And the *Water Mountain* or *Volcano* boils.

In Eq-ua-dor, then *Co-to-pax-i* scan;
As high o'er Chili flames the proud *Chil-lan*.

Hecla, in Iceland, and *Vesuvius* near
Naples, in Italy, the next appear.
Et'-na in Sicily, and the *Strom-bo-li*, (strom'-bo-lee')
Just north of Sicily, burns o'er the sea.

On the Canary Isles is *Ten-ne-riffe'*,
Fog-o on Cape-de-Verd rears her high cliff.
Ki-ra-uea on the Sandwich sits sublime,
And from its horrid crater pours forth slime.

Volcanoes.

More than two hundred volcanoes are known to exist in the world; one half of which are in America. But a great many have never been described, and have scarcely received a name.

Those of Europe and Asia are mostly on islands; while those of America are on the main land.

They are distributed as follows:

America,	on the continent, 97:	on islands, 19.
Europe,	on the continent, 1:	on islands, 12.
Asia,	on the continent, 8:	on islands, 58.
Africa,	unknown.	many.

More than forty volcanoes are continually burning between Cotopaxi and Cape Horn. Equador is one great volcanic district. Cotopaxi, Tunguragua, Antissana, and Pichinca, are the principal outlets for the internal fires.

The island of Java is noted as having a greater number of volcanoes than any other portion of the earth of the same size. A chain of mountains, in some parts 13,000 feet high, crosses the island, and, in the eastern part, is divided into a series of thirty-three separate volcanoes, most of which are in a high state of activity.

An eruption of one of the largest, in 1772, was one of the most terrible on record. The mountain, for a long time, was enveloped in a cloud of fire. Soon after, the immense mass sunk away, and disappeared, carrying with it ninety square miles of the surrounding country, forty villages, and three thousand inhabitants.

Kirauea, on Hawaii, one of the Sandwich islands, is another of the terrible volcanoes; its crater is seven and a half miles in circumference, and 1,000 feet deep.

View of Cotopaxi.

Cotopaxi is the loftiest volcano on the globe, and some of its eruptions have been the most tremendous. It is 18,890 feet high, and is one of the most beautiful summits of the Andes. It is a regular and smooth cone, wrapped in a vesture of eternal snow, which dazzles in the rays of the sun, with a superior splendor. Some of its eruptions have formed the most terrific and sublime scenes the eye ever witnessed. The flames have been known to ascend 3,000 feet above the top of the mountain. It is in a state of constant activity.

By a terrible eruption of Mt. Vesuvius in the year 79, the cities of Herculaneum and Pompeii were totally overwhelmed by the ashes and lava thrown from the crater of the volcano. These cities slumbered in silence beneath the congealed mass till the year 1750, when their sites were accidentally discovered by some peasants digging in a vineyard near the river Sarno. Since when, temples, theatres, shops, houses, paintings, &c., have been brought to light. Here skeletons were found, some in the attitude of prayer, some clasped together in each other's arms, and some with their treasures in their hands, as if trying to effect their escape.

"Of MAN here many a frightful form
In grinning horror stands,
Striving to 'scape the roaring storm,
His gold clenched in his hands.
Here skeletons by blood allied,
Locked in each other's arms,
Still lie embracing as they died,
In terror and alarms."

Mount Etna is one of the oldest volcanoes in the world, and has had some of the most terrible eruptions. One, in the year 1669, destroyed fourteen towns and 27,000 inhabitants. The lava thrown out formed a perfect river of fire, 1,800 feet wide, and 40 feet deep; and continued its course for more than 15 miles into the sea.

Mount Hecla is a celebrated volcano, on the island of Iceland. It is thirty miles from the ocean, and 5,530 feet high.

Skaptar Jokal, on the same island, had an eruption, in 1783, that ranks among the most terrible, in the destruction of life and the amount of lava thrown from its crater. No less than twenty villages, containing in all about 9,000 inhabitants, were consumed. It was estimated that the lava discharged would be sufficient to cover an area of 1,400 square miles, to the depth of 150 feet.

The *geysers*, or hot springs, or rather water volcanoes, spout hot water from 100 to 200 feet high, with a noise that resembles the discharge of a cannon.

OCEANS.

An *ocean* is a vast extent of brine,
Or salt sea water, boundless and sublime.

Five oceans there are found upon this ball:
Pacific, first, the largest of them all;
To Asia and America allied,
Eight thousand long, and full *twelve thousand* wide.

Atlantic, second, in the list survey,
Upon the west, bound by *America*;
While *Africa* and *Europe*, on the east,
Heave up their sea-walls to her waves of yeast;
Three thousand miles in width—*eight thousand* long,
In such a space the Atlantic sings her song.

The *Indian Ocean* is the third in size,—
Upon the north, the *Asiatic* shores arise;
Australia's east; while *Afric's* west her tide:
Four thousand long, and full *three thousand* wide.

The *Antarctic Ocean* laves the Southern Pole;
While, round the North, the *Arctic* billows roll:
Asia, and Europe, North America,
With Greenland, are the boundaries of this sea.

Three-fourths of the surface of the earth are covered with water, and the other fourth is covered by the land.

The water forms five great divisions, called OCEANS, viz. the Atlantic, Pacific, Arctic, Antarctic, and Indian Ocean. Beside these, there are many smaller divisions, called *seas*, *lakes*, *rivers*, &c.

The Pacific has an area equal to 78,000,000 square miles; the Atlantic, 20,000,000; the Indian Ocean, 12,000,000; the Antarctic, 10,000,000; the Northern, 2,000,000.

The extent of the different seas are as follows:—Chinese Sea, 1,000,000; Mediterranean, 8,000,000; Caribbean, 600,000; Okotsk, 500,000; Black Sea, 200,000; Red Sea, 100,000; Baltic, 9,000; Irish, 5,600.

The five great oceans form one continuous mass of water.

The Ocean is one of the sublimest works of Nature, whether it be in a state of rest, or aroused by storms.

Roll on, thou deep and dark blue ocean, roll,
Ten thousand fleets sweep over thee in vain,
Man marks the earth with ruin; his control
Stops with thy shore. Upon the watery plain,
The wrecks are all thy deeds.

* * * * * *

Thou glorious mirror! where the Almighty's face
Glasses itself in tempest, in all time,
Calm or convulsed, in breeze, or gale, or storm,—
Icing the Pole, or in the Torrid clime,
Dark-heaving, boundless, endless, and sublime,
The image of the Invisible! [*Childe Harold.*

LAKES.

Lake of the Woods, and Rainy Lake are found
  Skirting Columbia on her northern bound ;
Then comes *Superior, Huron,* and *St. Clair,*
  And *Erie Lake,* with one *Ontario* fair.
  'Tween Michigan and state Wisconsin roars
Lake Michigan, that laves the yankee shores.
  In Maine, is *Moosehead Lake* and *Um-ba-gog,*
With *Grand* and *Scoo'-dac* in the catalogue.
  And *Win-ni-pi-sio'-ge* on New Hampshire lain,
As 'tween Vermont and York is *Lake Champlain.*
  Oneida Lake, Cayuga, Seneca,
In New York state with Lake *O-was-co'* lay.
  Wisconsin hears her *Win-ne'-ba-go* talk,
With *St. Croix Lake, Flam-beau'* and *Tomahawk.*
Leech Lake, Itasca, Devil's and *Ottertail,*
  In Minnesota with *Fox Lake* we hail ;
Then *Pepin Lake* and *Spirit Lake* we see,
  And *Big Stone Lake* there finds a pedigree.
Salt Lake in Utah scours the Mormon border,
  Where *Utah Lake* rolls up in wild disorder.
  In California roars Lake *Bon'-ne-ville,*
There *Turtle Lakes* their rolling waves distil.
  In Mexico, *Tes-cu-co* and *Cha-pa'-la,*
As *Ni-car-a'-gua* lives in Guatamala.

  In Venezuela, *Maracaybo* view,
As *Tit-i-ca-ca* stands part in Peru.

  In Scotia stand *Loch Lomond* and *Loch Ness,*
With *Tay* and *Ran-noch* in their highland dress.
  Zu-rich, Lucerne and *Neuf-cha-tel* combine,
On Switzerland's mounts to feed a branch of Rhine.
Constance is north of Switzerland's rugged shore,
Geneva west, while south is Lake *Mag-giore'.*
  In Sweden, *Wenner, Wetter* and *Malar',*
Mid wild fantastic scenery take their fare.
In Russia, *Pe-i-pus'* and *Ill'-man* bide,
Where roars *O-nei'-ga* and *Lad-o'-ga* wide.

  Tsha-ny and *Baikal* in Siberia roar, (sha-ny)
Bal-kash is found upon the Tartar shore.
  As Afghan hears *Lake Durra's* wild harangue,
Ton-tia in China keeps with *Lake Po Yang.*

  Melgig and *Alshot Lakes,* are in Algiers,
  Dem-be-ah Lake in Tunis next appears.
  And last in Soudan, *Tchad Lake* finds a lair,
As Lake *Maravi* roars in Zanguebar.

Lakes are large bodies of fresh water, surrounded by land, which generally have an outlet into some ocean, gulf, or bay.

The great chain of lakes between the United States and British America discharge all their waters into the ocean, by the St. Lawrence river.

Lake Superior, the largest on the globe, stands at the head of this great chain. Its waters are elevated between 600 and 700 feet above the level of the Atlantic Ocean. It abounds with fish : trout, weighing from fifteen to fifty pounds, are caught in large quantities. The waters of this lake are remarkably clear,—a quality that pertains to all in this chain. The Pictured Rocks, on the southern shores, are great natural curiosities. They form a perpendicular wall of 300 feet, and extend from twelve to fifteen miles in length. The waters of this lake empty into Lake Huron, by the St. Mary's river.

Lake Baikal, in Siberia, is the largest body of fresh water on the eastern continent.

Ladoga and Oneiga are the largest in Europe.

Geneva, Neufchatel, and Lucerne, are elevated, among the Alps, more than 1,200 feet. They are distinguished for the wild, romantic character of their scenery, a feature that pertains to all lakes of mountainous districts ; such as those of Norway, Sweden, Finland, Scotland, Mexico, and South America. Their shores are usually lined with dark forests and rugged precipices.

The following table shows the size of most of the principal lakes.

European Lakes.

	Sq. miles.		Sq. miles.
Ladoga, Russia,	6,350	Constance, Switz.,	290
Wenner, Sweden,	2,150	Illmar., - -	275
Peipus, Russia, -	850	Maggiore, - -	150
Wetten, Sweden,	850	Neufchatel, - -	115
Malar, Sweden, -	760	Lucerne, - -	100
Geneva, Switzerland,	340	Garda, - -	180

Asiatic Lakes.

Aral, - -	9,930	Van, - -	1,960
Baikal, - -	7,540	Uroomiah, - -	760
Palkati, - -	3,696	Dead Sea, - -	500

African Lakes.

Lake Tchad, -	?	Dembea, - -	?
Maravi, - -	?	Dibbie, - - -	?

American Lakes.

Superior, - -	35,000	Arabasca, -	6,000
Huron, - -	20,000	Erie, - -	10,000
Great Bear Lake,	?	Ontario, -	7,200
Winnipeg, -	10,000	Titicaca, -	5,500
Slave Lake, -	12,000	Nicaragua, -	5,000
Michigan, -	16,000		

NIAGARA RIVER AND FALLS.

NIAGARA FALLS, AS SEEN FROM THE AMERICAN SIDE.

NATURE has many waterfalls, a few cataracts — ONE NIAGARA! That stands alone, vast, grand, indescribable! — the mighty alembic in which the world of waters is refined and etherealized! — the august throne upon which Nature sits, clothed in the glorious attributes of power and beauty! — the everlasting altar, at whose cloud-wrapt base the elements pay homage to Omnipotence! The floods that pour down its tremendous heights, seem gushing from the opened heavens, and plunging into the depths of the unfathomable abyss! Air groans, earth trembles, deep calleth unto deep, and answering thunders roll up the vast empyrean! Like a seething hell the gulf below sends up the smoke of its torment, and the foam of agony thickens upon the face of the dread profound, while far above upon the verge of the precipice, sits the sweet Iris — like faith upon a dying martyr's brow — arching the fearful chasm with its outspread arms, and smiling through all the terrors of the scene.

This cataract, the most wonderful and amazing curiosity in the natural world, is formed by the precipitous descent of the river Niagara down a ledge of rocks of more than one hundred and sixty feet perpendicular height, into an abyss or basin below, of unknown, but probably much greater depth. The river Niagara is that portion of the St. Lawrence, by which the lakes Erie and Ontario are united.

Some idea of the immense quantity of water forced

NOTE.—Most of the description here given of Niagara River and Falls, is taken from "PECK'S TOURIST'S COMPANION"; a work that should be in the possession of every true lover of Nature. The *language* and *descriptive talent* of the author, as will be seen by the few extracts here given, are in full keeping with the sublimity and beauty of his subject.

BRIDGE TO BATH ISLAND.

over the falls of Niagara, may be formed from the fact, that the lakes and tributaries which supply the river Niagara, cover a surface of not less than one hundred and fifty thousand square miles; and contain, as nearly as can be estimated, about one half of all the fresh water on the globe.

Niagara river is, in its whole course, quite in keeping with the stupendous cataract from which its principal interest is derived. There is nothing insignificant, nothing paltry, nothing common-place about it, from the lake in which its vast floods have birth, to that which they supply. It is every where grand, mighty, and majestic. When spread to the dimensions of a little sea, it has no resemblance to a shoal; and when contracted to the breadth of a creek, it seems to possess the power of an ocean. The very interruptions it meets with in its way, seem placed there only to exhibit the immensity of its force. The basin which receives its prodigious far-falling volume, resembles an abyss without bounds to its capacity; and the compressed channel through which it then flows, seems to have opened its rock-bound banks to an imprisoned sea, that would have burst a passage, had escape been denied.

Making a sharp angle at the Falls, it rolls on through beautiful curves, in an almost straight direction for about two miles; then winds gracefully off to the left, and passing through a succession of noble bends, rushes, wild, impetuous and uncontrollable, into the Whirlpool, where, like a baffled Titan struggling with his bonds, it rages and plunges round the impenetrable barriers that hem it in; and at last, having gathered anew its mighty energies, rushes headlong on in a fresh direction, and bounds away, free, fearless, and triumphant.

OUTLET OF THE WHIRLPOOL—CANADA SIDE.

Continuing in its new course—having turned less than a right-angle — but a short distance, it rolls away gradually to the west, and having gained its former direction, hurries on, inclining now to the right, and again bending to the left—here maddened by restraint, and there soothed by expansion, to the end of the mountain-plain, from the gasping jaws of which it rushes angrily forth, but soon recovering the serenity of its native seas, and no longer chafed or enraged, it flows quietly and smoothly on, through gentle curves and wooing banks, to the sweet lake whose soft embrace it has

Entrance to the Cave of the Winds.

Table Rock from below, as it appeared before its fall.

come so far, and encountered so much, to meet, and in whose peaceful bosom it finally sinks to repose.

The Cataract is made up of three distinct falls. The Great Horse-Shoe Fall is between Iris Island and the Canada shore. The Central Fall is between Iris Island and Luna Island. The American Fall is between Luna Island and the American shore.

The Cave of the Winds is back of, or behind the Central Fall. Reascend the sloping bank to the Central Fall, and the Cave of the Winds is before you. At the entrance, you pause to look up at the projecting cliff, and the sparkling torrent that shoots off far above, falling far over, and far below you ; and down at the piles of rock heaped up around, and the foam and the spray springing to light and loveliness from the rock-wave concussion. The mightiest throes give birth to the most beautiful things ; and thus the rainbow was born of the deluge.

You are on the steps descending into the cavern. The majesty, the sublimity of the scene cannot escape your notice, and you will feel what I find it impossible to express. A wall of rock rises frowning on one side ; the falling sheet arches the other. You see it leap from the cliff far above, and lash the rocks far below. You seem between two eternities, with a great mystery before you, whose secrets are about to be revealed. What a moment is this ! From the vast cavern into which you are passing, comes the sound of a thousand storms. You hear the mad winds raging around the walls of their imprisonment, and mingling their fearful roar with the reverberating thunders of the cataract ! The spray falls thick around you, and, almost overpowered with intense emotion, you hasten on, descend the steps, reach the bottom, instinctively retire from the rushing waters, and, having gained the centre and back of the cave, pause to look around. You seem all eyes, all ears, all soul ! You are in the sublime sanctuary of Nature—Her wonderful and fearful mysteries are above, beneath, and around you. God is Infinite, you are nothing ! This is His temple, you are His worshiper! It is impossible in such a place to be irreverent. The proudest, here is meek ; the haughtiest,

NIAGARA RIVER SUSPENSION BRIDGE.*

humble ; and the loftiest, lowly. The sights and sounds that crowd upon your gaze, and fill your ears, will be remembered to the latest day of your life ; nor will the emotions that swell your bosom and thrill your very soul, be ever forgotten.

THE SUSPENSION BRIDGE, two and a half miles below the Falls, spans the immense chasm of Niagara River, and serves as a connecting link between two great nations.

This stupenduous work was commenced in February, 1848, under the superintendence of CHARLES ELLET, jr., of Philadelphia, and finished during that year.

The length of this wonderful fabric, from tower to tower, is eight hundred feet. It is twelve feet wide, two hundred and thirty feet above the surface of the river, and capable of sustaining a weight of two hundred and fifty tons. It certainly is a triumph of art. There, over the raging element, it hangs, gorgeous and sublime, as a fit associate and companion of the mighty stream it crosses.

The immense wire cables, eight in number, that look like an inverted rainbow of faded colors; the strong towers over which they are suspended; the solid fastenings in the rocks at each end; the thick heavy planking that trembles at the lightest breeze, and undulates 'neath our footsteps,—combined with the sullen roar of the savage stream beneath us, and the giddy, painful height to which we are suspended, inspire us with the highest emotions of awe and sublimity.

There is another bridge of much greater length, though of less elevation, now in process of building, six miles and a half below this, at Lewiston, designed to connect this village with Queenston, on the Canada side. Its length between towers is some fourteen hundred feet, and will form, when completed, another great highway between Canada and the United States.

* From Burke's Guide to Niagara Falls.

MINNESOTA.

VIEW OF THE FALLS OF ST. ANTHONY.

THIS TERRITORY is bounded on the North by Canada West, on the East by Lake Superior and Wisconsin, on the South by Iowa, and on the West by the Rivers Missouri and White Earth, having an area of 160,000 square miles. It comprises all that portion of country situated at the head waters of Mississippi, Lake Superior, and many of the northern branches of the Missouri River. The face of the country is no where broken by mountain chains, although many portions are highly elevated, consisting of immense *plateau* or table-land, which sends out, from inexhaustible reservoirs, some of the largest streams on the face of the globe. But the greater part of this country, consists of rolling prairie, oak openings, with forests of pine, tamarack, beech, and the sugar-maple. In these immense uncultivated districts, are found all kinds of wild game : there is the bear, the fox, the large grey wolf, the deer, and the antelope. Also, the wild goose, the duck, and prairie hen. Pigeons in vast swarms likewise abound in these forests.

No country in the world has a greater number of rivers, lakes, and springs, than Minnesota. Besides the Mississippi and its innumerable branches, here the swolen flood of the Missouri finds a supply. Here the mighty St. Lawrence, with its wide-spread lakes, has its origin. And from these regions, Winnipeg, Lake of the Woods, and Rainy Lake draw their waters. These waters are well stocked with fish, that furnish the wild Indian and adventurous pioneer with food.

The lands are all well adapted to agriculture ; barley, oats, wheat, and potatoes, are produced in abundance. The strawberry, raspberry, blackberry, and blueberry grow spontaneously, of a large size and an excellent quality.

Travelers, visiting this Territory, all speak in the most enthusiastic terms, of its picturesque scenery, of its lovely lakes, sparkling and cool springs, its falls and cascades, its healthy and bracing climate, and of the strange superstitions of the untutored Red Man that still roams over its wildernesses.

THE FALLS OF ST. ANTHONY, rank as first among the curiosities of this Territory. The Mississippi River at the falls, is 627 yards wide, and is divided into two unequal channels by Cataract Island, which extends several rods above and below the Falls, having a width of about one hundred yards.

The view on page 71 represents the Eastern channel, as it appeared in a state of nature. A *dam* has recently been thrown across to Cataract Island, so that the beauty of the fall is destroyed.

The fall of the Western channel has met with a still worse fate. The whole limestone rock, over which the waters poured in one unbroken sheet, and behind which travelers walked in safety, has lately broken away, so that the waters now run down an inclined plane, instead of driving over a precipice.

The fall of water, in either channel, is not more than 20 or 25 feet, and is sublime, only when taken in connection with the rough, savage scenery around.

St. Paul, the capital, and largest town in the Territory, is situated on the North or left bank of the Mississippi, 8 miles below the Falls of St. Anthony, and 327 miles by water from Galena in Illinois. It has a population of over 1200 inhabitants, and is doubtless destined to be a large and important city.

St. Anthony, at the Falls, is situated on the East side of the river, and is fast advancing in population and importance. It has an excellent water power, healthy location, and will, doubtless, be a place of fashionable resort.

Mendota, three miles above St. Paul, on the opposite side, is a small though important village, from its being at the mouth of the St. Peters River.

Fort Snelling, directly across the St. Peters, from Mendota, is situated on a high bluff. The Military Reservation here, embraces about 100 square miles.

Kaposia, an Indian village on the West bank of the Mississippi, and five miles below St. Paul, has a population of 300 souls.

Stillwater, at the head of Lake St. Croix, is 18 miles by stage from St. Paul. Its population is about 1000.

Pembina, about the size of Stillwater, is situated on Red River in latitude 49°. It is the most northern town in the Territory.

Pilot Knob, 262 feet above low water, in the Mississippi, is a commanding eminence two miles from the mouth of St. Peters River.

Maiden's Rock, or Lover's Leap, is a high promontory, on the East side of Lake Pepin. Here a beautiful Indian maiden, it is said, being compelled to marry against her will, threw herself down upon the rocks beneath, and was picked up a lifeless corse.

Fountain Cave, so called from a rivulet of pure water that flows through it, is situated near the bank of the Mississippi, two and a half miles above St. Paul. It is 150 feet long, 20 wide, and composed of white sand-stone, resembling sugar-loaf.

Painted Rock, two miles above Stillwater, on the St. Croix River, is a high precipice, on the face of which are, carved and painted, numerous images, figures, and hieroglyphics. This place is held in much veneration by the Indians.

CALIFORNIA.

CALIFORNIA, as ceded to the United States by Mexico, is bounded by Oregon on the North, the Rocky Mountains on the East, Mexico on the South, and the Pacific Ocean on the West; comprising an area of 400,000 square miles.

THE STATE OF CALIFORNIA embraces nearly one half of this territory, or about 180,000 square miles; consisting of a large extent of land, bordering on the Pacific for 800 miles, having a uniform width of about 230 miles; and reaching from Oregon on the North, to Mexico on the South.

THE GOLD REGION is that portion which lies in the Valleys of the Sacramento and San Joaquin Rivers; where a greater quantity and abundance of the precious metal has been procured, than in any other part of the known world. The gold is found here in its virgin state, in three distinct deposits; 1st—in the sand and gravel beds; 2d—mixed with decomposed granite rock; and, 3d—mixed with *talcose* slate.

THE CLIMATE of California varies very much in different parts. In the Valley of San Juan, it is said to be that of a paradise, mild, healthy, and serene. While in the Valleys of Sacramento and San Joaquin, it is subject to great extremes of heat and cold. The year is divided into two seasons—the *wet* and the *dry*. The former commences in December and continues till March; the latter lasts during the remainder of the year.

Whether or not the soil and climate of California are adapted to agricultural purposes, is a question of much controversy. Wilkes gives it as his opinion, that the amount of arable land in this portion of California, will not exceed 12,000 square miles; though, by the process of irrigation, he thinks it would prove exceedingly productive.

As to the salubrity of the Climate, it may justly be remarked, that in no part of the world, could men expose themselves so much to the hardships of toil and deprivation, and suffer less from the effects, than in California.

The population at the present time, numbers as high as 300,000 ; and it is made up of the most heterogeneous and motley mass of human beings, of any other country under heaven. Here, every language is spoken; and here, after a lapse of thousands of years, the noise and confusion of Babel is again heard. Here are men of every profession and trade—of every rank and condition in life. Here are rich and poor, learned and unlearned; and, contrary to every other country, the true *nobility* are those that *dig the dirt.* And if this noble democratic principle will last, then California will have produced a corner-stone for the monument of true democracy, that will do more to commemorate her fame than the golden block taken from the Sierra Nievada, to adorn the monument of Columbia's honored Son.

The only good harbors of California, remarks Wilkes, are San Diego, San Francisco, and Bodega. There are besides several roadsteads, which have been used as anchorages during the summer season, viz.: the Bays of Monterey, San Pedro, and Santa Barbara. San Diego is the most Southern port in the State, and is believed, by Bayard Taylor, to be the best on the Pacific coast, with the exception of Accapulco, in Mexico.

Bodega lies to the North of San Francisco ninety miles, and is of less importance than either of the others.

From California Sacramento 's rolled,
Southward and west, through regions rich with gold.
To Sacramento drives San Jo-a-quin,
Its course north westward on the map is seen.

San Francisco, the largest and by far the most important town in the State, is situated on a bay of the same name. In a commercial point of view, this city bids fair of becoming the emporium of the trade, the commerce, and the wealth, of the vast Pacific. With a railroad connecting it with the Atlantic, it would doubtless take rank among the first cities of the globe.

Sacramento City, the second town in size and importance, is situated on the Sacramento River; by an inundation of which, it was once nearly destroyed. A levee has recently been thrown up, for the purpose of protecting it from further encroachments.

Sutter, named in honor of Capt. Sutter, and situated at the head of navigation on the Sacramento, is noted as the point from which the roads issue leading to the Northern mines.

Vernon, at the junction of the Sacramento and Feather Rivers, is a town of some importance from the country around it.

Stockton, the commercial depot for the Southern mines, is situated on the waters of the San Joaquin.

Alvero, at the head of the Bay of San Francisco, is situated in one of the most fertile districts of California.

Among other towns of note, may be mentioned San Jose, the present Capital of the State, and situated near the Southern extremity of the Bay of San Francisco.

UTAH, or the district of the Great Salt Lake, called also Deseret, is situated West of the Rocky Mountains, on the direct line of the great Overland Route to California. It is near the shores of this lake, that the Mormons have established themselves. From this place to the Gold Regions, the journey requires forty-five days, with wagons.

Salt Lake, the waters of which are very salt and bitter, is about seventy miles long, and from forty to sixty wide; being elevated some 5,000 feet above the level of the sea. Some of the lands around this lake are extremely fertile, while others are barren. Bear River, emptying in from the North, is exceedingly cold and transparent.

Utah Lake, the waters of which are fresh, empties into Salt Lake, by the channel of the Utah River.

The entire basin of the Great Salt Lake, or, more properly, the Territory of Utah, comprises an extent of land measuring 160,000 square miles; some portions of which are fertile and productive. but the greater part is composed of dry, arid, plains, the rivers of which have no outlet, and either evaporate in their course, or empty into dead salt lakes, or, more properly, pools of stagnant water covered over with a yellow skum, or saline incrustation.

The country situated East of the Colorado, North of the Gili, and West of the Anahuac Mountains, comprises a vast extent of 120,000 square miles. It is very imperfectly known, but is represented by travelers that have passed through it, as consisting mostly of high table-lands, dry, barren, aud unproductive—many of the streams of which, including the Colorado, are salt and bitter.

THE

RULES OF ARITHMETIC.

IN VERSE.

Addition.

Addition, is joining more numbers than one,
And putting together to make a whole sum,
Addition's the rule that learns us to count,
And the sum that 's produced is called the *amount.*

RULE.

The numbers write down, as the rule comprehends,
Placing units under units, and tens under tens;
Draw a line underneath, and commence at the right,
Or the unit column, the work to unite;
If its sum or amount should not exceed 9,
Then place it direct 'neath its own native line:
But if 9 it exceeds, then the unit you place
'Neath the column of units, (the units to grace);
While the tens or the figure that's to the left hand,
To the next column join, as you well understand.
Observe the same rule, till you come to the last,
And the whole amount write as this column you cast.

Subtraction.

Subtraction, it teaches, when numbers are given,
One greater, one less, as 10 stands to 7,
To find out their difference, for difference we see,
And when worked and achieved, we find to be 3.

RULE.

The numbers first write, the less under the greater,
Placing units and tens, in lines of their nature,—
The subtrahend, then, from the minuend take,
And that which remains, an answer will make.—
But if in the less number, a figure we find,
Which exceeds that above it, let 10 then be joined
To the figure above, and from the amount,
Take the figure below, (nor mistake in the count),
But forget not to add, to the next figure, then
In the subtrahend, *one* to make up for this *ten.*

Addition.

ADDITION is joining together two or more numbers, to make one whole sum or amount.

Addition is the rule by which we count, or put numbers together.

The whole sum, or answer, is called the *amount.*

RULE.

Write down the numbers, one under the other, placing units under units, tens under tens, and draw a line underneath.

Begin at the right hand, or unit column, to add or unite the numbers together; add together all the figures contained in that column.

If the sum or amount should not exceed 9, then place it under the column; but if it does exceed 9, put the right hand figure under the column, and carry the left hand figure, and add it on to the next column.

Observe the same rule, putting down under the column added, the right hand figure, if it exceeds 9; and carrying the left hand figure to the next column.

At the last column, write down the whole amount, and the work is complete.

Subtraction.

SUBTRACTION is taking a less number from a greater to find out the difference, as 7 from 10; the difference, or remainder, is 3.

The greater number, or the number to be lessened, is called the *minuend.* The less number, or the one to be taken from the greater, is called the *subtrahend.* The difference, or that which is left after the operation of the work, is called the *remainder.*

RULE.

Write down the numbers, the less under the greater, placing units under units, tens under tens, and draw a line underneath.

Subtract the less from the greater: commence at the right hand figure in the lower line, and take it from the one above it in the upper line; write the difference below the line. So proceed till the whole is subtracted.

If the figure above should be less than the one below, then add *ten* to the one above, and from the amount, take the figure below. But in this case you must add one to the next left hand figure, in the lower column. This is called borrowing ten.

Multiplication.

Now, *Multiplication*, its nature I 'll show,
It 's a short way of working *Addition*, you know,
When the same number comes, in prose or in rhymes,
To be used or repeated, a number of times—
Let the *less* number under the *greater* one stand,
Call one the *multiplier*, one the *multiplicand*,—
Name the answer the *product*,—and then just annex
For the sign of the rule, the letter—X

RULE.

First, the number *above*, must be multiplied o'er
In succession, by each figure found in the *lower*,
While the same as Addition, the rule you have seen,
Remember to carry *one* for every *ten;*
While the *right hand* figure of each product must lie
Direct 'neath the figure you multiply by;
Then the same as *Addition* their products unite,
And the amount of them all is the answer quite.
Or when the multiplier is 100 or 10,
Or 1, with any number of ciphers, I mean,
Of ciphers, annex to the multiplicand,
As many, as in the multiplier stand.
Or when ciphers are in the multiplier found,
Or between the significant figures abound,
By *figures significant* only, perform,
While the *right* of each *product* directly is borne
'Neath the figure you multiply by. (Now retain
This rule forever secure in your brain).

Multiplication.

MULTIPLICATION is a short way of performing Addition, when the same number is to be repeated a number of times.

The number we multiply by, is called the *multiplier.*

The number to be multiplied, is called the *multiplicand.*

The answer is called the *product.*

The sign of Multiplication is the letter X.

RULE.

When the multiplier exceeds 12.

Write down the multiplicand, under which, write the multiplier, placing units under units, tens under tens, and draw a line underneath.

Multiply the multiplicand by each figure of the multiplier, commencing at the right hand; and remember to set the first product of each figure directly under the figure in the multiplier by which you multiply.

Add these several products together, and the amount is the product required.

To multiply by 10, 100, 1000, *&c.*

Add to the multiplicand as many ciphers as there are ciphers in the multiplier; and the multiplying is performed.

When ciphers occur between the significant figures of the multiplier, we omit them, multiplying by the significant figures only, minding to write the first product of each figure, directly under the figure by which we multiply.

To prove multiplication, divide the product by the multiplier, and if the quotient is the same as the multiplicand, the work is right.

Division.

Next simple *Division*, the fourth Rule is seen,
It 's a short way of working Subtraction, (I ween),
It shows us Subtraction, its smallest remains,
And how often one number another contains.
The *Divisor* is that, which divides, as you see,
The *Dividend's* that, which divided must be.
The *answer* is called the *Quotient*, and shows
How oft the divisor in the dividend goes.

RULE.

Write the dividend down, and to the left hand,
With a curve line between, the divisor must stand,—
Then of figures, as many divide, (and consign)
As will hold the divisor, times not over nine, (9)
With the number arising, the quotient supply,
Which by the divisor you then multiply,—
The product then take from the dividend o'er it,
And beside what remains, the next figure lower it;
Which again you divide, if 't will hold the *divisor*,
If not, in the quotient a cipher we tie sir,

Division.

DIVISION is a short way of performing many Subtractions; or,

It shows how often one number is contained in another.

The Dividend is the number to be divided.

The Divisor is the number that divides the dividend.

The answer is called the Quotient, and shows how often the Divisor goes into the Dividend.

RULE.

When the Divisor is more than 12.

Place the Divisor at the left of the Dividend, separated by a line.

Then assume as many figures of the dividend as will hold the divisor something less than 10 times.

See how often the divisor is contained in the assumed portion of the dividend, and place the result at the right of the dividend, separated by another line.

Multiply the divisor by this figure, and place the product under the part assumed or divided, and subtract it therefrom, and to the remainder bring down the next figure for a new dividend.

And to our *remainder*, a figure once more,
From the dividend bring, and proceed as before.

WHEN THE DIVISOR IS LESS THAN 12.

But when the divisor does not exceed twelve,
By *short division* the problem we solve,
'Neath the dividend then the quotient you bind,
While the process is mostly performed in the mind.

Reduction.

Reduction is changing a kind and its name,
To another, and keeping its value the same.
It consists of two kinds, Ascending is one,
Descending the other, by which we come down;
In Reduction ascending, division we try;
In Reduction Descending, we then multiply.

Reduction Ascending.

Divide the lowest kind that stands in your sum,
By that number it takes of the sum to make one
Of the next higher order, and keep the same round
'Till the problem is solved, and the answer is found.

Decimal Fractions.

In decimal Fractions, your work is the same,
As when in whole numbers, the problems you frame.

Addition and Subtraction of Decimals.

RULE.

In Addition of Decimals, Subtraction too,
The same as whole numbers, the work you must do;
Write tenths under tenths, and hundredths, likewise,
You place under hundredths, the rule to comprise.
Let the decimal point, if the work you approve,
Fall precisely 'neath those in the numbers above.

Multiplication of Decimals.

TO POINT OFF IN MULTIPLICATION OF DECIMALS.

If in *Multiplication* of *Decimals*, then
Point off from your product, with pencil or pen,
For *decimal places*, as many as stand
In both *multiplier* and *multiplicand*.
If the product *in figures deficient* is found,
To the left of the product let ciphers be bound.

Division of Decimals.

TO POINT OFF IN DIVISION OF DECIMALS.

In *Division of Decimals*, then you may count
From the right of the *quotient* the whole amount

Divide this the same as before, and to the remainder continue to bring down figures from the dividend till the whole is divided.

To prove Division, multiply the divisor and quotient together, and if the product is the same as the dividend, the work is right.

Example;—2840÷40=71, the Quotient. To prove this, multiply 71 by 40, thus: 71×40=2840, the same as the dividend.

Reduction.

REDUCTION is changing one kind or denomination to that of another, without altering its value.

It is of two kinds: Reduction Ascending and Descending; the former is performed by division, and the latter by multiplication.

RULE FOR REDUCTION ASCENDING.

Divide the lowest denomination given, by as many as it takes of the same to make one of the next highest order.

Divide the quotient in the same manner, by the number it takes of its own denomination to make one of the next higher denomination; so continue to do till it is reduced to the denomination required.

Decimal Fractions.

DECIMALS are performed the same as whole numbers. The only difficulty is to know where to put the separation or decimal point, between decimals and whole numbers.

Addition and Subtraction of Decimals.

Write down the numbers, one under the other, placing those of the same value under each other; or, units under units, tens under tens, &c. Likewise, tenths under tenths, hundredths under hundredths, and then add or subtract as in addition or subtraction of simple or whole numbers.

Let the decimal point in the *sum*, or *remainder*, fall directly under those in the sum.

Multiplication of Decimals.

To point off in Multiplication of Decimals.

Multiply the same as in whole numbers, and point off in the product, for decimal places, as many figures as there are decimal places in both multiplier and multiplicand, counted together.

To multiply a whole number by a decimal, the product is less than the *multiplicand*; for example, ,5 multiplied by ,5 the product is ,25.

Division of Decimals.

To point off in Division of Decimals;

Divide the same as in whole numbers, and point off from the right of the quotient, for decimals, as many places as the decimal places in the dividend

That the *dividend* numbers o'er the *divisor*
In *decimal figures*—and if the supply (sir)
In the quotient, of figures, deficient you find,
To the left of the *quotient let ciphers be joined.*

exceed those of the divisor; and if there be a deficiency of figures in the quotient, supply such deficiency by annexing figures to the left of the quotient.

To divide a whole number by a decimal, the quotient is greater than the *dividend*: for example, 250, divided by ,5, the quotient is 500.

Interest.

Interest is a certain per cent. that's allowed,
For the use of money on the lender bestowed.
The *principal* 's that, which is loaned or *lent,*
The rate, on each dollar, is called the *per cent.*—
It is *Simple* and *Compound*—The rule for the *first*
When desired for one year, may thus be rehearsed:

RULE.

First, the *principal multiply* by the *rate per cent.*
And divide by 100 the product, (attent)
If for more years than one, the product it bears
Must be multiplied by the number of years.
If the interest for months, in your sum is implied,
By 12, the interest of one year, divide,
And the *quotient* by the *number of months multiplied,*
The interest in full, for the months will decide.
If the use of your money for days you would see,
The amount for one month by 30 must be
Divided, and then the quotient you raise
Be multiplied o'er by the number of days;
Add the *days* and the *months* and the *years all* in one,
And the answer desired will be the whole sum.

Interest.

INTEREST is a per cent. paid by the borrower to the lender, for the use of money.

The sum of money loaned or lent, is called the *principal.*

The per cent. is the annual amount paid, as so many dollars for the use of a hundred.

RULE FOR SIMPLE INTEREST.

Multiply the *principal* by the *rate per cent.*, and divide the product by one hundred, and the quotient is the interest for one year.

Multiply this last by the number of years, and the product is the interest for the years.

To compute the interest for months;

Divide the interest of one year by 12, and the quotient is the interest for one month; multiply this by the number of months, and the product is the interest for the months.

To compute the interest for days;

Divide the interest of one month by 30, the number of days in a month, and the quotient is the interest for one day.

Multiply the interest of one day by the number of days, and the product is the interest for the days.

Add the days, months, and years together, and the amount is the interest required.

Compound Interest.

Now interest *Compound,* to you I will show,
'Tis *interest on interest* and *principal* too,
Which are added together as interest is due.

RULE.

First find the amount for one year, the same
As in simple interest, the rule that you've seen,
Then this is the principal for the next year,
Which again you compute with patience and care.
And again to the product the interest unite,
Which becomes for the third year, a principal quite.
So continue, and from the amount of the last,
Subtract the sum loaned, and the interest is cast.

Compound Interest.

COMPOUND INTEREST, is interest on interest, where the interest is added to the principal at the end of each year, as it becomes due.

RULE.

First find the amount for one year, and this amount is the principal for the second year.

Then perform, with this principal, the same as with the first, finding the amount for the second year, which amount is the principal for the third year; so continue to do, finding the amount for each year, and from the last amount, subtract the sum loaned, and the remainder is the Compound Interest for the number of years required.

Rule of Three.

RULE.

Of the three *given numbers,* a *third term* you make
That 's of the same kind with the *answer you seek;*
And then just consider the *question* in hand,
Whether greater or less, the answer will stand

Rule of Three.

Of the three given numbers, make that the third term which is of the same kind with the answer sought.

Then consider, from the nature of the question, whether the answer will be greater or less than the third term.

Than this the third term,—If *greater* 't is known
That of the two numbers the *greater* comes down
For the term that is second, or term number two;
While the less number's *first*, as the pencil will show.
But if smaller your answer than term number *three*,
Reverse the *two terms*, let the *less second* be,
Then the second and third you next multiply,
And divide by the *first* and the answer is nigh.

Alligation.

Alligation is mingling or mixing together,
Teas, sugars or spirits (and one thing or other),
It divides itself thus, (now be sure and learn it),
Alligation Medial, Alligation Alternate.

Alligation Medial.

Alligation Medial is finding the *mean*,
The middle or average 'twixt either extreme
Of several simples, some less and some greater;
So read o'er these lines, and they 'll learn you its nature.

RULE.

Supposing a merchant has three kinds of tea,
At 10 shillings, 5 shillings, and shillings 3,
Which he wishes to mix and together confound,
And then wants to know what's the worth of a pound,
Add your 10 and your 5 and your 3 as you mix,
And divided by 3, the quotient is 6.

Six shillings per pound, price of the mixture.

Alligation Alternate.

Alligation Alternate is the rule that finds,
What quantity of any number of simples or kinds,
Whose rates are all given, direct as we state,
To compose a mixture of a specified rate.

RULE.

Arrange in a column your rates for command,
And place the *mean rate* off at the *left hand*,
Each rate that is *less* than the *middle* or *mean*,
Join with one that is greater, as is plain to be seen,
Place the difference 'tween each rate and mean kind,
Opposite that with which it is joined.

Square Root.

RULE.

Divide into *periods* of *two figures* each,
The number you know, as the pedagogues teach,—
In the *left hand period* find the greatest *square*,
Which from it subtract, and to what remains there
Bring the next period down for a *Dividend* (fair):
Place the root of the square at the right hand of all,
And two times the root a *Divisor* we call.

If greater, place the greater of the two remaining numbers for the second term.

If less, place the lesser of the remaining numbers, for the second term.

In either case, multiply the second and third terms together, and divide by the first term; and the quotient will be the fourth term, or answer.

Alligation.

ALLIGATION is mixing together several simples of different qualities, or prices, so that the composition may be of some intermediate quality or price.

It is of two kinds, Alligation Alternate, and Alligation Medial.

Alligation Medial.

ALLIGATION MEDIAL, is finding the mean or average proportion or price, of several numbers or prices.

RULE.

Add together the several prices or ingredients, and divide the amount by the number of ingredients.

Or when there are a greater number than one of each kind,

Multiply the number by the price, set the products in a column, add the several products together, and divide the amount by the amount of the several ingredients, and the quotient is the mean price of the composition.

Alligation Alternate.

ALLIGATION ALTERNATE teaches to find what quantity of any number of simples, whose rates are all given, will compose a mixture of any specified rate.

RULE.

Arrange the rates of the simples in a column under each other, with the mean price at the left hand.

Connect each rate, that is less than the mean rate, with one or more that is greater; place the difference between each rate and mean price opposite that with which it is joined, and it will be the quantity required.

Square Root.

RULE.

Divide your number into periods of two figures each, by putting a point over the unit figure, and every second figure from the place of units.

Find the greatest square in the left hand period, and put the result in the root, at the right of the number.

Square this figure, and place the square under

Then try the Divisor, see how many times
The Dividend holds it (by prose or by rhymes).
Of its *right hand figure* exclusive, you know,
And write in the root the number 't will go,
Then to the *Divisor the same figure* tie,
And by the same figure the whole multiply;
The *product* then *take* from the Dividend (penned),
And of that which remains, make a new dividend;
By bringing the *period* that' s next, *along side*,—
And for a *Divisor* that 's new and untried,
Just *double* the *figures* that stand in the *root*,
And work as before, till the answer is got.

the left hand period. Then subtract it therefrom, and to the remainder bring down the next period for a dividend.

Double the root, already found, for a divisor, or see how many times it is contained in the dividend, exclusive of its right hand figure, and place the result in the root, for the second figure of it, and likewise put the same figure at the right hand of the divisor.

Multiply the divisor with the last figure annexed, by the last placed in the root, and subtract the product from the dividend, and to the remainder bring down the next period for a new dividend.

Double the figures already found in the root, for a new divisor, and from these find the next figure in the root, as last directed, and so proceed till the whole is finished.

Cube Root.

RULE.

Your number divide, as I shall prescribe,
In periods of three figures each, side by side,
In the left hand *period* the greatest cube find,
Put its *root* in the quotient, and then you must mind
To subtract from the period, the Cube that is found,
And by what *remains*, the next period bring down
For a *dividend*,—then a *divisor* to spy,
By 300 your *quotient's square* multiply;
Then as *Simple Division*, the work you perform,
But *subtract not* the product—let this be forborne.
Then the square of the last quotient figure espied,
By the *first* quotient figure, must be multiplied,
And the answer arising by 30 be tried (or *multiplied*),
And the product of these placed under the last,
That units and tens in their lines may be cast.
Write the cube of the last quotient sign, under all,
And the amount of the whole, a subtrahend call,
Which you must subtract from the dividend *o'er* it,
And *by what remains* the next *period lower* it
For a new dividend, with which you proceed
As before, till the *root* in the *quotient* you read.

Cube Root.

RULE.

Separate the given numbers into periods of three figures each, by putting a point over the unit figure, and every 3d figure beyond the place of units.

Find the greatest cube in the left hand period, and set the root in the quotient.

Subtract the cube, thus found, from the said period, and to the remainder bring the next period down for a dividend.

Multiply the square of the quotient by 300, calling it the divisor.

Seek how many times the divisor may be had in the dividend, and place the result in the root; then multiply the divisor by this quotient figure, and write the product under the dividend.

Multiply the square of this quotient figure by the former figure or figures of the root, and this product by 30, and place the product under the last; under all, write the cube of this quotient figure, and subtract the amount from the dividend, and to the remainder bring down the next period for a new dividend, with which proceed as before, until the work is finished.

Geometrical Progression.

The first term, ratio, and number of terms being given, to find the last term.

A few leading powers of the ratio write down,
With each index placed o'er, beginning at *one*,
The *indices* whose *sum* as the rule thus informs,
Shall approach within *one* of the number of terms,
Stand over the factors, whose product must be
Multiplied by the *first* term, and the last term we see.

Geometrical Progression.

RULE.

First put down a few leading powers of the ratio, with the indices placed over them, beginning at *one*. Add the most convenient indices together, to make an index one less than the number of the term sought.

Multiply together the powers belonging to these indices, and their product, multiplied by the first term, will be the answer.

THE END.

AGENTS WANTED.

AGENTS ARE WANTED TO SELL THE POETICAL GEOGRAPHY IN NEARLY EVERY TOWN IN NEW ENGLAND. As no work of the age meets with more rapid sales, it presents a rare opportunity for active enterprising men to make money.

For an Agency in any part of the New England States, apply to

L. W. TURNER,
MERIDEN, NEW HAVEN CO., CONN.

RECOMMENDATIONS.

The Poetical Geography.—This is a very meritorious and interesting work; and in regard to its originality is equalled by few, and surpassed by none of the works of the day. * * If it does not succeed and become a popular one, it certainly is not the *Author's* fault.—*N. Y. True Sun.*

It is written by George Van Waters, who has laboriously and ingeniously thrown the whole science of Geography into rhyme. * * * * The lines flow smoothly, and give evidence of ease and skill in versification. It will prove as effectual in fastening the principal facts of Geography upon the memory, as the common verse of "Thirty days hath September, &c.," is in fixing the days of the month.—*N. Y. Evening Post.*

This is a book for every body, admirably calculated and fitted for the wants of the million. It is founded upon a known law of the mind; and discovers no mean effort on the part of its author to bring into easy reading verse the almost unpronounceable names of Woodbridge's Classified School Atlas.—*Wis. Waukesha Freeman.*

This is an original composition—a model cut from a new quarry, bearing upon its features novelty and usefulness combined. It is evidently a work of labor, and one that must have highly taxed the ingenuity of the rhymer. We think it can not fail of answering the purpose for which it was designed, viz: of giving the leading features of Geography a lasting stamp upon the memory. But aside from the poetry, there is prose enough in the work to amply pay its extreme cheap price.—*Cleve. True Dem.*

"A BOOK THAT IS A BOOK."

The Poetical Geography, by George Van Waters.—Too much cannot be said in favor of this interesting publication. It is certainly a curiosity in literature. A composition that must have required a patient, plodding brain to have produced. It forms a complete system of Geography, containing all the capes, rivers, cities, towns, islands, mountains, &c., of the globe, put into smooth poetry, giving the locations of places, and for what they are noted and distinguished. Aside from the music of its *quaint rhymes*, its leading quality appears to be its *multum in parvo:* its comprehensiveness; so much told, using so few words—and certainly one important point in literature is to strip it of its redundancies. The prose part seems the quintessence of the sublime and wonderful in nature and art, as imposing natural scenery, mountain peaks, volcanoes, cataracts, celebrated fabrics, ruins, antiquities, &c.

The object of the work, is to aid the memory—on the principle that nothing is more easy to learn and remember than poetry, while, on the other hand, few things are more difficult than hard names and locations.—*Cleve. Plain Dealer.*

It is expected that Agents will act the honorable part with subscribers, and never sell the work for less than the subscription price, for this is abusing our patrons, which is the *worst species of ingratitude.*

As the Poetical Geography is sold only by subscription, Agents are expected to call at every house, that every one may have a chance to purchase one or more copies.

www.ingramcontent.com/pod-product-compliance
Lightning Source LLC
LaVergne TN
LVHW011120110826
845150LV00008B/2206
9781425503864